BIOL 250L

MICRO BIOLOGY Lab

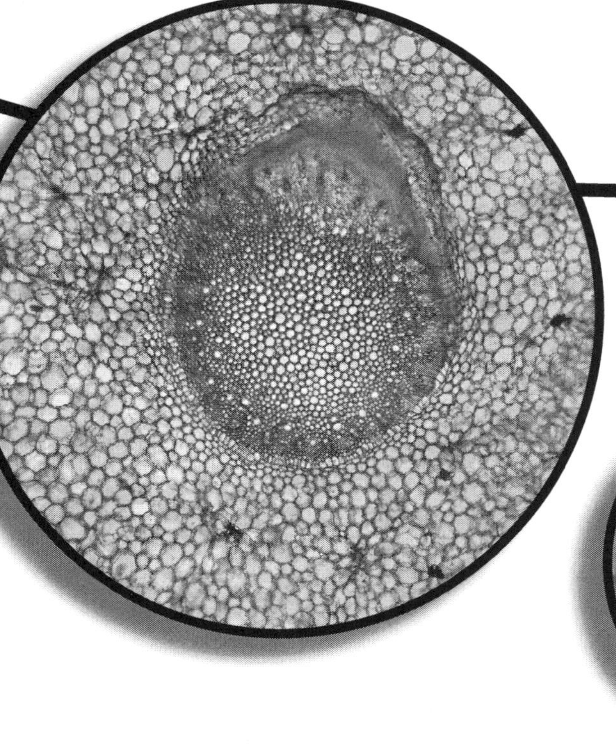

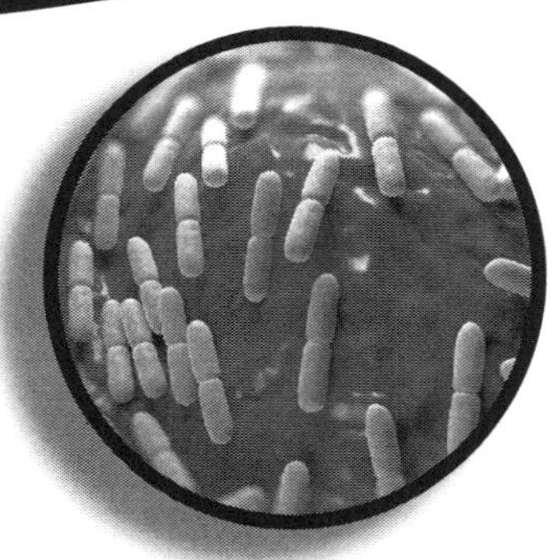

Second Edition
University of South Carolina, Columbia
Department of Biological Sciences
Isaac M. Hagenbuch, PhD
Contributing Author Shannon Pipes

Microbiology Lab

BIOL 250L
Second Edition
University of South Carolina, Columbia, Department of Biological Sciences
Isaac M. Hagenbuch, PhD
Contributing Author: Shannon Pipes

Printed in the United States of America
10 9 8 7 6 5 4 3 2 1
ISBN: 978-1-61740-688-1

Van-Griner Learning
Cincinnati, Ohio
www.van-griner.com

President: Dreis Van Landuyt
Project Manager: Maria Walterbusch
Customer Care Lead: Lauren Houseworth

Hagenbuch 688-1 F19
316215
Copyright © 2021

Table of Contents

Factors That Influence Microbial Growth

Assignments

Microbiology Laboratory Rules

Violation of these rules will result in forfeiture of points.

1 | Attendance is required as mandated by the University of South Carolina policy. ***All*** lab sessions (classes and return sessions) are to be attended by ***all*** students. If you are late to lab, you cannot take the quiz for that class, and there will be no make-ups. Absence due to a contagious illness should be documented by a medical professional, and any concession is at the discretion of the lab instructor.

2 | **Cheating, which includes but is not limited to plagiarism and falsification, will not be tolerated.** ***All*** work must be your own (even if data are collected as a group) — all lab reports must be your own thoughts and written in your own words. Do not copy another student's work, and do not copy out of the lab manual. There is a zero-tolerance policy in the Department of Biological Sciences towards plagiarism. ***Plagiarism = Failure. All*** incidents of cheating will be reported to the Office of Academic Integrity.

USC Statement of Academic Integrity

Academic ethical behavior is essential for an institution dedicated to the promotion of knowledge and learning. The University of South Carolina is committed to fostering a university environment which exemplifies the values embodied in the *Carolinian Creed*. All members of the University Community have a responsibility to uphold and maintain the highest standards of integrity in study, research, instruction, and evaluation as well as adhering to the *Honor Code*.

University of South Carolina Honor Code
The *Honor Code* is a set of principles established by the university to promote honesty and integrity in all aspects of a student's academic career. It is the responsibility of every student at the University of South Carolina to adhere steadfastly to truthfulness and to avoid dishonesty in connection with any academic program. A student who violates, or assists another in violating the *Honor Code,* will be subject to university sanctions. I understand that attendance is required and that any failure to abide by the *Carolinian Creed* and *Honor Code* will result in a loss of points. Name: ____ Date: _________________

Rules Regarding Safe & Appropriate Laboratory Conduct

1 | You will at some points in this lab be working with potential human pathogens. Therefore, consider all microorganisms you work with in this lab to be potential pathogens and handle them safely. The Microbiology Laboratory rules are for your protection and the protection of people *outside* the lab to whom you may carry microbial pathogens.

2 | No smoking/vaping, eating, drinking water bottles, or snack foods in the lab. If you need a drink or a snack, wash your hands and partake of it outside of the lab. Do not chew gum, the ends of pens, your fingernails, etc. Your mouth is a wide-open portal of entry.

3 | Glass pipettes require the use of a pipetter. ***No mouth pipetting*** is permitted.

4 | Always wear shoes with closed toes. If you do not, you will be asked to leave the lab and change your shoes to ones with closed toes.

5 | Long hair must be tied back because we work with flames.

6 | **Wash your hands thoroughly (10–15 sec) with soap and warm water before starting work and before leaving the laboratory (even if you are just visiting the restroom).**

7 | *Never under any circumstances* take cultures out of the laboratory.

8 | **Notify the instructor *immediately* in the case of personal accidents — *no matter how small*.** These include cuts, burns, spills, any introduction of microbial material into the eyes, mouth, or broken skin, etc.

9 | Swab the bench top with 5% Lysol at the beginning and end of each laboratory period. If a culture is spilled, cover the spill with a Lysol-soaked paper towel and notify the instructor.

10 | Long pants are required in lab to cover and protect your exposed lower half. Natural-fiber clothing you don't mind permanently staining is an acceptable substitute. The stains used to stain cells in this laboratory are ***permanent dyes*** and cannot be fully removed from clothing.

11 | Gloves should be used when handling live cultures or stains. Skin is generally a sufficient barrier for all microbes we use in this lab, but gloves provide an additional measure of safety to both you and those outside of the lab.

12 | Use the shelves at the front of the lab to store coats, hats, backpacks, books, etc. All such things, except your lab notebook and manual, must be kept off of the benchtop. Your possessions can become contaminated by simple contact. Anything upon which a culture is spilled must be autoclaved to rid it of living microbes.

13 | ***Do not use the sinks for waste disposal*** unless otherwise directed by your instructor. This includes disposal of cultures.

14 | Worn gloves, used solid media, disposable inoculation instruments, and culture spill clean-up materials must be discarded into the labeled biohazard containers. ***Under no circumstances*** **will you discard these items in the standard trash cans.**

15 | All other items such as paper towels, parafilm wrappers, scraps of paper, inkless pens, exhausted pencils, etc. will be discarded in the black trash bins located at the end of each bench and by the exit.

16 | When collecting media for the day's exercises, take only the number of tubes and plates specified. Identify each medium by marking with a Sharpie.

17 | **All plates and tubes must be properly disposed of in the specified containers after data are recorded.** Test tubes are to be labeled with your Sharpie and the labels must be removed (using ethanol) from *all* test tubes before they are put in racks for sterilizing.

18 | Glass pipets and micropipette tips must be placed in designated containers after use.

19 | Never open an agar plate that shows signs of fungal growth. Keep the fungal spores contained.

20 | **Students should wear safety glasses when working with a flame.**

21 | Keep alcohol away from any flame. You will be required to work with alcohol and flames during the course of lab. Caution is to be used whenever you are working with a flame but *especially* when alcohol is involved.

22 | Bunsen burner flames must be kept low. Never let them exceed 3 inches in height.

23 | Microscopes are numbered and will be inspected. They must be returned *clean and to their proper space.* All oil must be removed from the objectives using lens paper.

24 | The syllabus is a contract, which can only be modified by the TA or Lab Coordinator. Any modifications will only occur with sufficient notice.

25 | Expect that labs will always take the full allotted time — **don't ask if we'll end early.**

26 | Turn cell phones to silent or vibrate. **Cell-phone disturbances will *not* be tolerated.** Text/tweet/snapchat/facebook/tinder at your own risk. Failure to pay attention in lab can get you hurt and will get you a failing grade.

27 | **Late work policy:** Lose half a letter grade (5%) per day late. **e.g.:** B+ would be dropped to a B.

28 | If you have a question about anything, ask the TA.

29 | It is **your** responsibility to prepare for labs and **your** responsibility to listen to directions in lab.

30 | Clean up before you leave. Don't leave a mess or a disorganized space for those who come after you.

31 | Class disturbance is unacceptable. If you are a distraction to others or present behavior problems, you will first be asked to desist. If you continue, you *will* be asked to leave the lab. If you refuse, USC police will be called to remove you.

32 | If you arrive to lab short of sleep and show evidence of this, you may be asked to leave. Tired people are a hazard to themselves and others, especially in a lab with fire and infectious organisms.

I have read rules 1–32 and agree to abide by them while in lab.

Name: ____   Date: _______________

Laboratory 1
Aseptic Technique & Culture Transfer

Microorganisms intended for laboratory and scientific investigation are grown and kept in **pure cultures.** It is a regrettable reality that microorganisms are not visible to the unaided human eye. One consequence of this is that technicians have no immediately visible way of determining if they are keeping their microbial cultures pure. Therefore, the first technique that any student of microbiology must learn is **aseptic technique.**

Aseptic technique is a set of best practices and procedures designed to

- facilitate the creation of pure microbial cultures;

- keep microbial cultures pure and uncontaminated;

- keep the microbes only in certain, specific places and at specific times; and

- keep people outside the lab safe by not transferring any microbes from the lab into the larger world.

Every microbial culture we use is many times more concentrated than what you would find in the natural world. This means that they all present a certain amount of danger to those who handle them. There are so many microbial cells present that exposing cuts, mouths, ears, eyes, etc. directly to them can result in an infection that will require medical treatment. Properly executed aseptic technique and adherence to the laboratory safety rules will ensure that you reduce your risk level to the lowest level possible.

Governing Principles of Aseptic Technique

- **Fire as a sterilizing tool.** Inoculation loops and needles will be sterilized by heating until they glow orange in a Bunsen burner flame (Figures 1-1 and 1-2). Once sterile, loops/needles must be cooled sufficiently before contacting any microorganisms to avoid killing them. Glass rods will be stored in ethanol and residual ethanol burned off before the rod is cooled and used. Glass cools slower than metal.

- **Use of adductive vertical air currents.** When working with open nutrient media, whether inoculated or not, work within the envelope of adductive (moving toward center) and vertical (upward) air currents created by the Bunsen burner flame (Figure 1-3). This will keep dust particles and their attendant microorganisms from settling into your plates. Lightly flame the necks of test tubes before re-capping to create a convection current that will carry air *out* of the tube.

- **Limiting exposure of nutrient media.** Media must be capped or lidded at all times unless you are transferring organisms or material in or out of the container. Tubes of medium should be held at an angle and not vertically. Caps and lids must not be placed on the benchtops.

- **Keeping silent and attentive while inoculating media.** Accidents happen when you are inattentive, therefore keep your attention on your work. Also, do not talk while making organism transfers as saliva-containing aerosols coming out of your mouth can contaminate your medium or your transfer instrument. Do not blow onto inoculating instruments to cool them!

- **Avoiding the creation of bacteria-containing aerosols.** Always cool your inoculation, transfer, or spreading instrument to a temperature that won't kill the bacterium nor boil the medium in which they reside. Boiling, shaking, sonication, and other agitation methods can produce bacteria-containing aerosols that can lead to contamination not only of your nutrient medium but of your lab bench, skin, clothing, or lungs.

- **Storing and incubating solid medium plates inverted.** Always incubate and store your solid media plates in an inverted position to prevent any condensation that may appear from flowing onto the medium surface. Dust sticks to water droplets. Therefore, any of this sort of water/medium contact may contaminate your media either before or after inoculation.

- **Disinfecting work surfaces.** Surfaces, such as table tops, should be disinfected with 70% ethanol (EtOH) or Lysol before and after each use. This serves to keep the number of potentially contaminating microorganisms to a minimum. Even having done this, do not set down things like inoculation loops, lids, or caps unless it is absolutely essential (such as to deal with a safety issue). If you must place a cap on a benchtop, always place it with its inner surface facing downward.

Procedure

1 | Collect an **Eppendorf tube** containing live bacterial culture.

2 | Obtain two additional tubes of sterile **tryptic soy broth (TSB)** and label them (A & B, 1 & 2, whatever convention you want to use).

3 | Light your Bunsen burner and adjust it until you have a well-defined consistent flame (Figures 1-1 and 1-2).

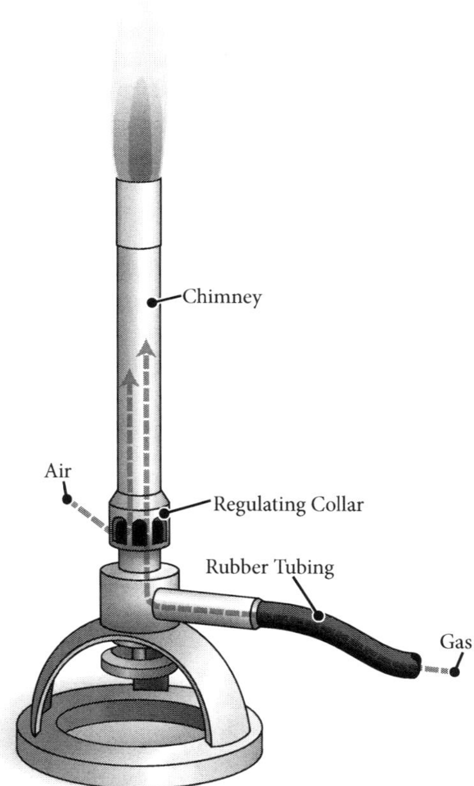

Figure 1-1 | Bunsen Burner. A Bunsen burner is a simple natural gas jet. Its main parts are the metal chimney and the air intake regulator. The chimney allows the gas and air to mix before burning at the top of the chimney. More air = hotter flame, less air = cooler flame. Balance the air and gas to get the type of flame you need.

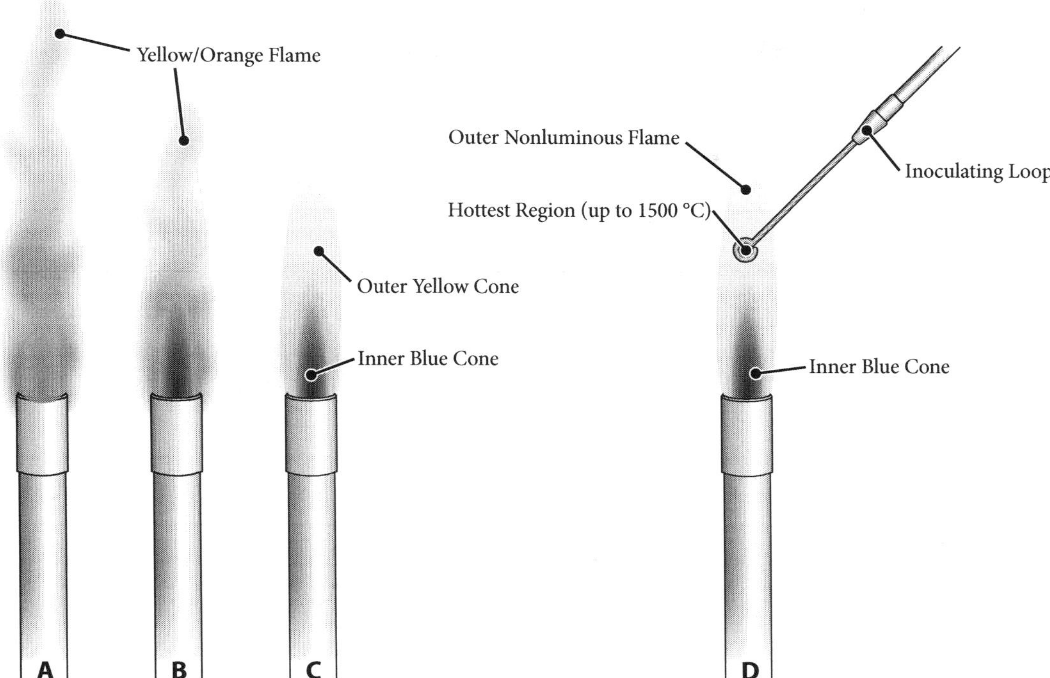

Figure 1-2 | Bunsen Burner Flames. Here are some examples of the types of flames you get with different gas/air mixtures. The general rule is that a yellow/orange flame is burning inefficiently (not hot enough). A) Is a result of too little air and is not useful for sterilizing because it is too cool. B) Is closer to the ideal flame but there is still too little air/too much gas. C) Is useful for sterilizing but not yet the most efficient. D) Is the ideal where there is no indication of inefficient burning and there is a bright blue flame, indicating the hottest possible fire.

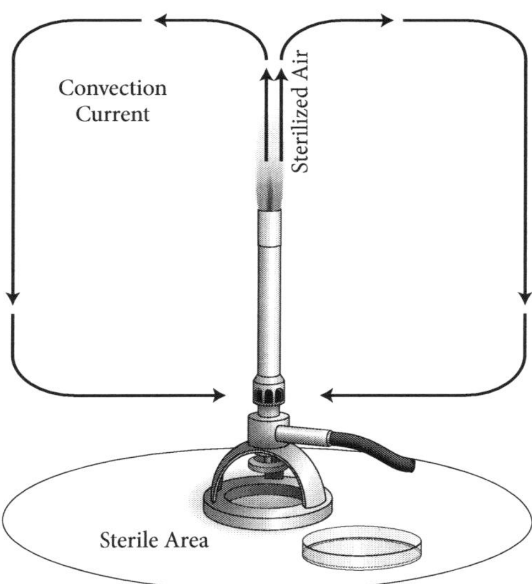

Figure 1-3 | Sterile Area. Hot air rises and draws the air below it upward. On a tabletop, this results in a 3D donut of convection currents surrounding a Bunsen burner flame. The lateral and upward movement of the air will direct dust particles and aerosols away from your medium surface.

4 | Sterilize your inoculating loop within the flame by heating the wire until it glows orange-red.

5 | Remove your loop from the flame and cool it (count to 10). Do not blow on it! Your breath can contain droplets of saliva that carry bacteria which will contaminate your inoculation instrument.

6 | Use your inoculating loop to collect a sample from the live culture and transfer it to one of your tubes of TSB.

7 | Flame your loop again to sterilize it. Wire must glow orange-red to be sure of sterilization!

8 | Cool the loop and place it in the *other* tube of TSB.

9 | Remove the loop and flame it again before putting it aside.

10 | Incubate the TSB tubes for 24 hrs at 37 °C.

11 | After incubation, you should have a live bacterial culture (cloudy) in your first tube, and the second tube should still be sterile (clear).

Laboratory 2
Microbes of the Air

Every type of microbe may be found in the air at various times and in various places. Airborne organisms are transient. There is no known microorganism that carries out any major part of their lifecycle preferentially while airborne. Within building interiors fungi are generally the most numerous **culturable microorganism** in the air followed by bacteria.

We are able to culture ~1% of the microbes that we have genomic evidence for. Clearly, we have much work yet to do in this area. Though we have had limited success so far, the creation and maintenance of microbial cultures is a key skill in both scientific and clinical pursuits. Many factors contribute to the successful culturing of any microorganism, including the following:

- Culture medium
- Temperature
- Moisture
- Gasses

You should therefore not expect that any medium will allow you to culture *all* the microorganisms that are present in a given environment. In this experiment, you will be using **tryptic soy agar (TSA)** to determine if there are culturable microbes in the air of the laboratory. TSA may allow the growth of some fungi, but it will also allow the growth of a wider array of bacteria, if present.

Procedure

1 | Obtain a TSA plate and label it with your name and section.

2 | Open the lid of the plate and leave the medium exposed for some period of time (note how long).

3 | Close the lid and **parafilm** it (see Figure 2-1).

4 | Incubate at room temperature for up to one week.

5 | If, after incubation, the plate has anything fuzzy growing in it, ***do not open it.*** That fuzziness is likely a fungus that is rife with fungal spores. If you open it, you risk spreading spores hither and yon.

6 | Choose a bacterial colony to describe and describe it using proper cultural-characteristics terms.

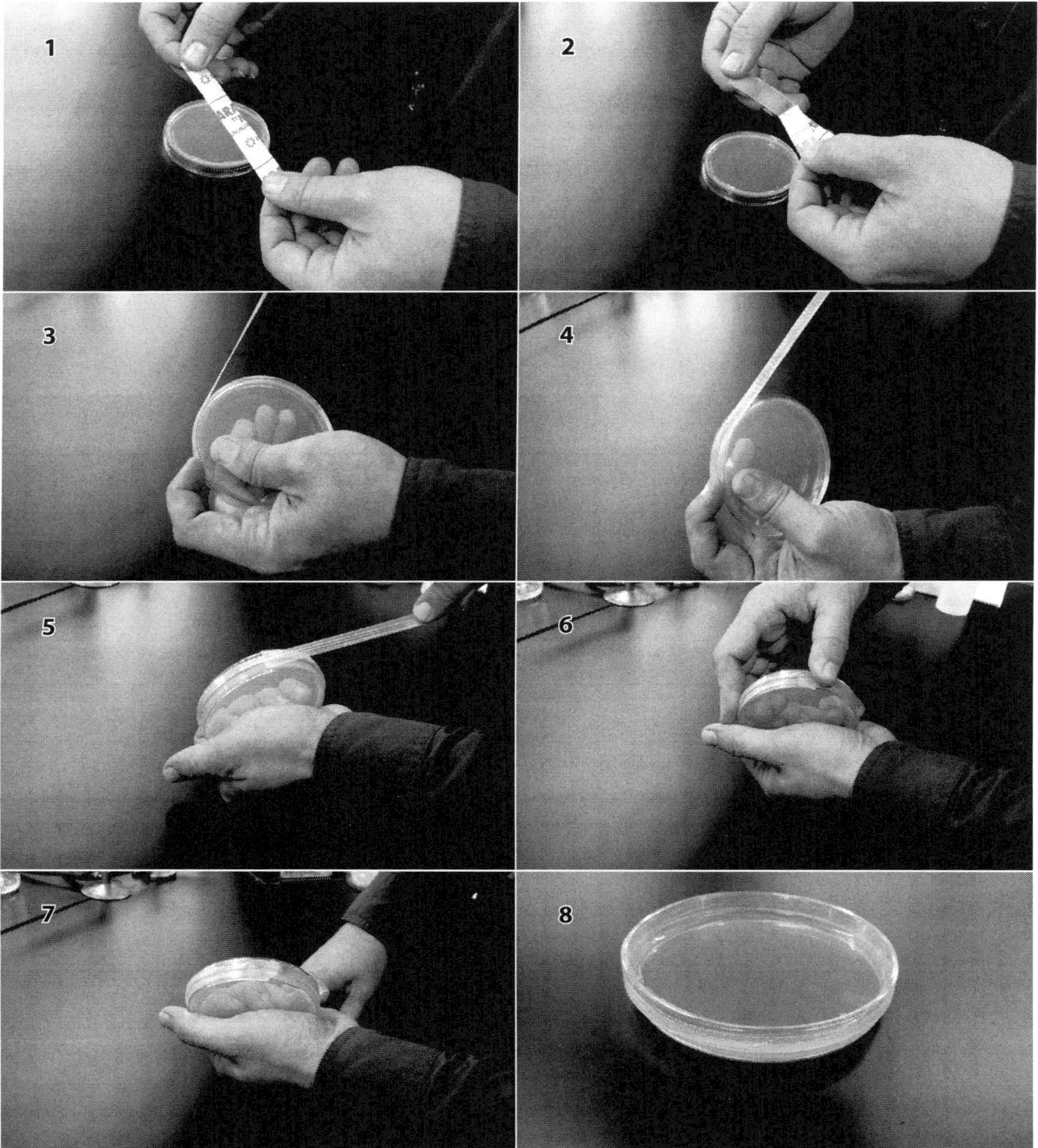

Figure 2-1 | How to Parafilm Your Petri Dish. Parafilm will adhere to itself once stretched out. Separate the Parafilm from its backing and stretch it completely around your Petri dish and firmly attach the end. Be careful that you cover the entire gap between the lid and base. Inspect your Parafilm job for holes before placing it in the incubator. Holes in your parafilm can result in a dried-out, ruined culture.

3 Laboratory 3
Sampling Environmental Surfaces for Microbial Presence

You may have been told that microbes are everywhere, but do you know that for certain? Are the same number and types of microorganisms present on any given environmental surface? These are questions you will begin to answer by the end of this laboratory exercise.

The same caveats that apply to culturing microbes from the air also apply here. There is no way to culture *all* microorganisms that may be present on environmental surfaces. Your results will therefore only allow you to say something about whether there are culturable microorganisms present.

If for no other reason, then at least for the sake of your own self-respect, don't choose some basic location to sample. What is a basic location? Anything within arms-reach as you sit in your chair. The lab benchtop, drawer handles, the inner handle of the microlab door, the lab faucet handles, etc. Take your sampling materials and leave the lab room. Go find someplace that may actually be interesting. Floors, walls, and windows are generally cold and don't have enough direct human contact to carry anything.

Procedure

1 | Obtain a tube of sterile saline, a sterile swab, and a TSA plate.

2 | Find a non-basic surface to test.

3 | Wet the swab with the saline. Recap but keep it in your hand.

4 | Take your moist swab and rub it all over your chosen sampling location. Be sure to press firmly and to rotate the swab as you sample to get an accurate representation of the organisms present.

5 | Pop the saline cap again and submerge the swab in the saline. Close the lid on the swab stick to brace it and then *break the handle off* leaving the swab inside the container.

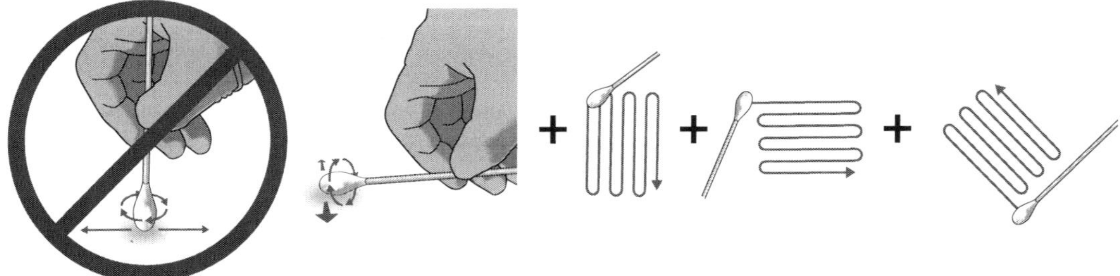

Figure 3-1 | Surface Sampling. Rub vertically, then horizontally, then diagonally as shown. Rotate the swab whilst rubbing.

Figure 3-2 | Crosshatch Patterns That Can Be Used on Solid Media

6 | Return to lab and find a free vortexer. Vortex your swab container for 10–20 seconds. Make sure the liquid is actually spinning within the container.

7 | Bring your swab back to your station and use your pipette to create a spread plate with 250 µl of the liquid.

    a | Remove a glass spreader from the ethanol beaker and light the alcohol on fire with the Bunsen burner and let the alcohol burn off. ***Do not hold the glass in the Bunsen burner flame.***

    b | Let the spreader cool (count to 30 or more).

    c | Place your plate on a turntable or lab bench, open the lid (do not set it down!), place the glass spreader flat against the medium surface, and rotate the plate as you use the spreader to spread an even layer of the sample over the medium surface.

    d | Replace the lid and parafilm it shut.

8 | Discard the swab container into the regular trash.

9 | Incubate the plate at room temperature for up to one week.

10 | After incubation, if you find anything fuzzy growing in the plate, ***do not open it.***

11 | Observe a bacterial colony and describe its colony characteristics.

Subculturing in Broth & on Solid Media

The transfer of live, pure microbial cultures from one place to another and without contamination is an indispensable skill for anyone who may need to collect or preserve microbial specimens in a professional/clinical context. When transferring microbes, intentional transfer is called **inoculation** and inadvertent transfer is called **contamination.** Generally, we want to maximize our ability to inoculate while simultaneously minimizing contamination. Toward this end, you will employ aseptic technique as you perform inoculations to and from liquid and solid microbiological media.

When streaking on a solid medium, the goal is to obtain well-separated (discrete) colonies. The streaking technique is a type of dilution. You are diluting the culture by spreading it over a greater and greater surface area. You want to spread the culture out enough that you are able to separate individual cells far enough apart that they can give rise to separate colonies composed of millions of cells just like the original cell. These are what we call pure colonies, and we must be able to consistently obtain them so we can be sure that we are maintaining the purity of our microbial cultures.

Procedure

1 | Obtain for yourself:

    a | One tube of bacterial culture.

    b | One Petri dish (plate) of sterile tryptic soy agar (TSA).

2 | Being careful to observe aseptic technique:

    a | Sterilize your inoculation loop.

    b | Transfer live bacteria from the culture onto the plate using a standard streaking technique.

    c | Sterilize your inoculation loop.

    d | Transfer live bacteria to the TSA plate while successfully completing a standard streaking technique and obtaining a goodly amount of discrete colonies.

Your laboratory instructor will demonstrate the proper technique, and you may also refer to Figure 4-1.

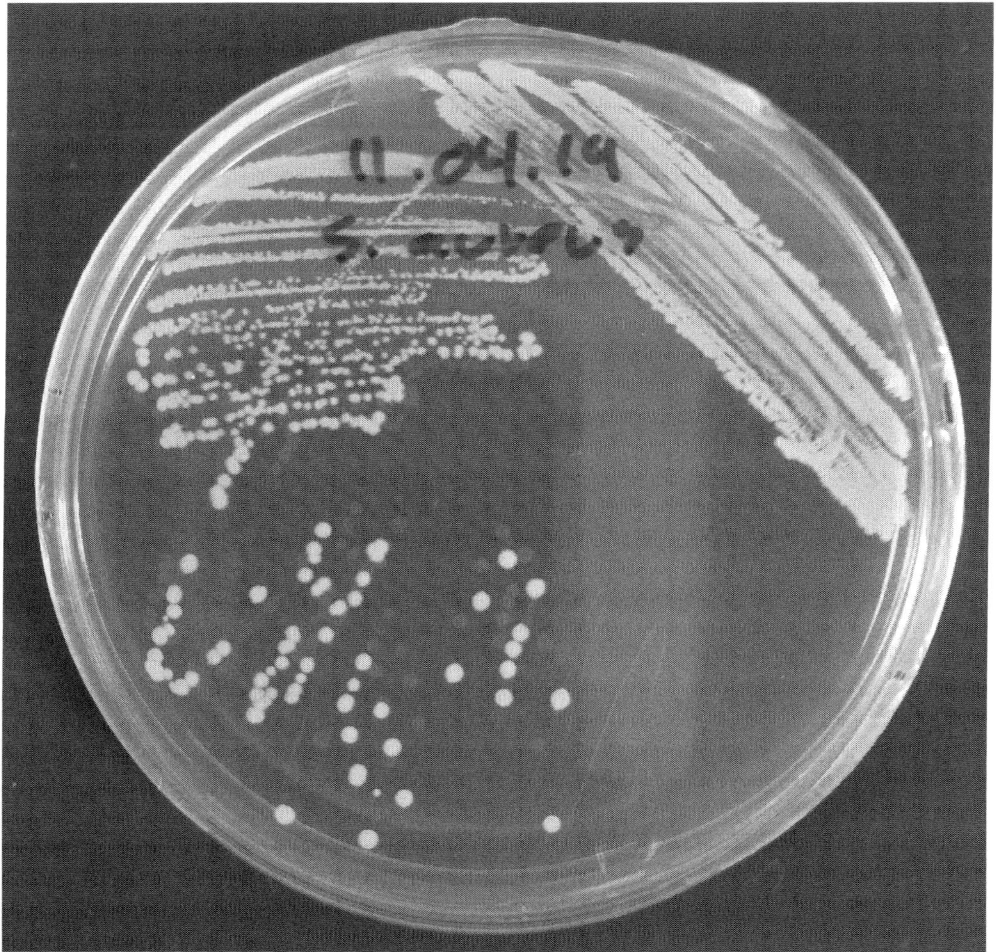

Figure 4-1 | 3-Way Streak Technique.

Laboratory 5
Cultural Characteristics of Microorganisms

The appearance of microbial cultures is often useful in identification of microbes. Differences in appearance are manifested in a number of characteristics that can vary according to species, but it is important to keep your culture conditions constant. Why? Because cultural conditions (temperature, nutrient profile, medium additives) can cause a single species to show variation. The studious person can find cultural characteristics for most known bacterial species within the volumes of *Bergey's Manual of Systematic Bacteriology,* but a basic knowledge is sufficient to be useful in most contexts.

This laboratory exercise will give you the terms you need to use when describing colonies from your air exposure plate, environmental swab plate, and in various exercises through the rest of the semester.

NOTE: If any of your plates have filamentous (fuzzy) colonies, ***do not open the plate.*** These colonies are likely fungi, and opening the plate may release fungal spores into the air of lab. Fungal spores are already a common contamination issue, so let us not make the issue worse.

Choose a colony in your air and surface plate and apply some combination of the following **cultural characteristic** terms to a description of them.

Cultural Characteristic Terms

Nutrient Agar Plates

This is perhaps the most commonly used type of microbiological medium. It is very useful for maintaining the purity of cultures. These are often inoculated using 3- or 4-way streak technique. Discrete colonies on solid media may differ in the following ways:

1 | **Size** – Pinpoint, small, medium, large.

2 | **Pigmentation** – Colony color. Some organisms are **chromogenic** and will therefore produce pigments that color the colony or the medium upon which it grows. Most organisms are non-chromogenic and therefore appear a shade of white, gray, or tan.

3 | **Form** – The shape of the colony which may be one of four basic types.

    a | **Circular** – A smooth, unbroken periphery.

    b | **Irregular** – An edge with a non-uniform indentation.

    c | **Rhizoid** – Growth that spreads in a root-like way.

    d | **Spindle** – Ovoid, or eye-shaped growth with a swelled middle and pinched ends.

4 | **Margin** – The appearance of the outer, leading edge of the colony.

    a | **Entire** – Even (not bumpy), sharply defined.

    b | **Lobate** – Marked, irregular indentations.

    c | **Undulate** – A wavy appearance.

    d | **Erose** – Serrate. Regular, sawtooth-like.

    e | **Filamentous** – Fuzzy, thread-like.

5 | **Elevation** – The difference in height between the surface of the medium and the colony surface.

   a | **Flat** – No discernable difference in elevation.

   b | **Raised** – Colony surface slightly above medium surface.

   c | **Convex** – Hemispherical or dome-like.

   d | **Pulvinate** – Convex but with a higher dome elevation and a resulting larger volume.

   e | **Umbonate** – A two-step elevation consisting of a raised portion with a convex portion on top of it.

The same organism will generally yield additional information if grown in/on other types of media. While we don't use these as often in the lab, it is worth having a familiarity of the terms associated with them.

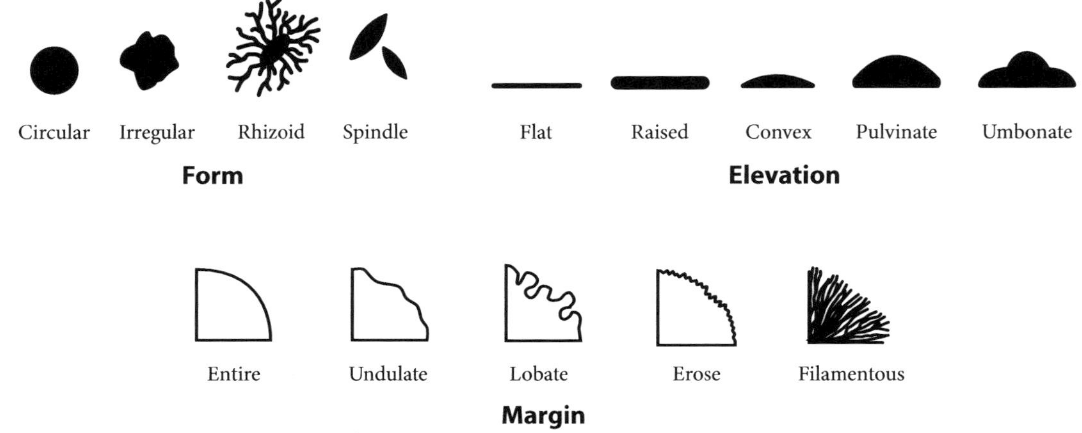

Figure 5-1 | Cultural Characteristics for Colonies on Solid Media

Nutrient Broth

Here, you are looking for the distribution and clarity of growth.

1 | **Uniform/fine turbidity** – A non-particulate, finely-dispersed growth throughout the medium.

2 | **Flocculent** – Flaky or dust-bunny-like aggregates throughout the medium.

3 | **Pellicle** – A firm, pad-like growth covering the surface.

4 | **Sediment** – Most of the visible cells are congregated at the bottom of the culture vessel. Often flocculent or grainy.

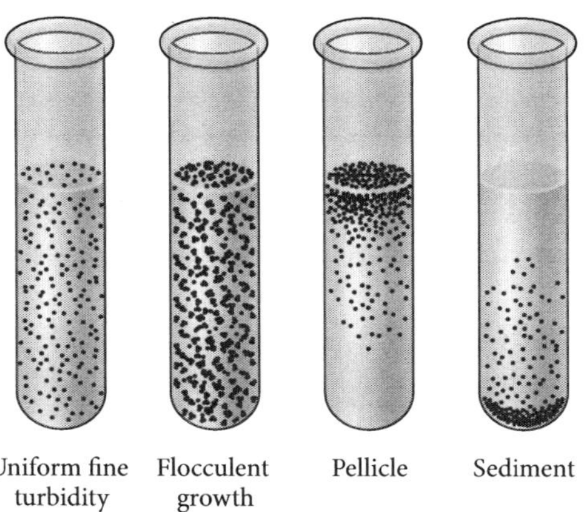

Figure 5-2 | Cultural Characteristics for Broth Cultures

Nutrient Agar Slant

These use the same medium you find in nutrient agar plates, but in a test tube set at an angle as the liquid agar cools. These are inoculated with a single, straight line.

1 | **Abundance of growth** – None, slight, moderate, heavy.

2 | **Pigmentation** – Color of the colonies.

3 | **Optical properties** – Determined by how much light travels through the colony growth: Transparent (all light makes it through), translucent (some diffuse light makes it through), or opaque (no light makes it through).

4 | **Form** – The appearance of that single-streak inoculation.

    a | **Filiform** – Continuous growth with smooth edges.

    b | **Echinulate** – Continuous growth with irregular edges.

    c | **Beaded** – Non-continuous, semi-continuous growth.

    d | **Effuse** – Thin, spreading growth.

    e | **Arborescent** – Treelike.

    f | **Rhizoid** – Rootlike.

5 | **Consistency** – This has to do with how firm or moist the colonies are.

    a | **Dry** – Apparently free of moisture.

    b | **Buttery** – Moist and shiny like room-temperature butter.

    c | **Mucoid** – Glistening, shiny, slimy.

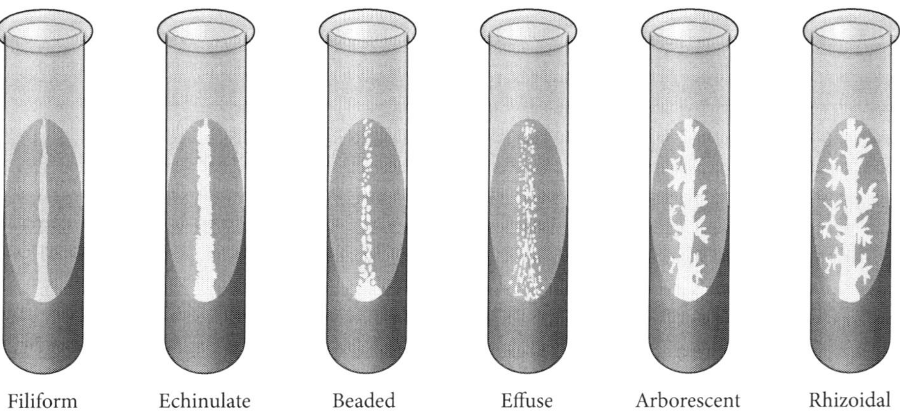

Figure 5-3 | Cultural Characteristics for Colonies on Agar Slants

Nutrient Gelatin

This medium is commonly used when identifying unknown organisms because the ability to liquefy hardened gelatin is not common to all organisms. This medium is inoculated by stabbing once. Take care to be consistent because the liquefaction pattern is influenced by depth of inoculation.

1 | **Crateriform** – Saucer-shaped liquefied surface.

2 | **Napiform** – A bulb–shaped liquefaction at/just under the service.

3 | **Infundibuliform** – Funnel-shaped.

4 | **Saccate** – Tubular, extending down through the medium.

5 | **Stratiform** – Complete liquefaction of the top half of the medium.

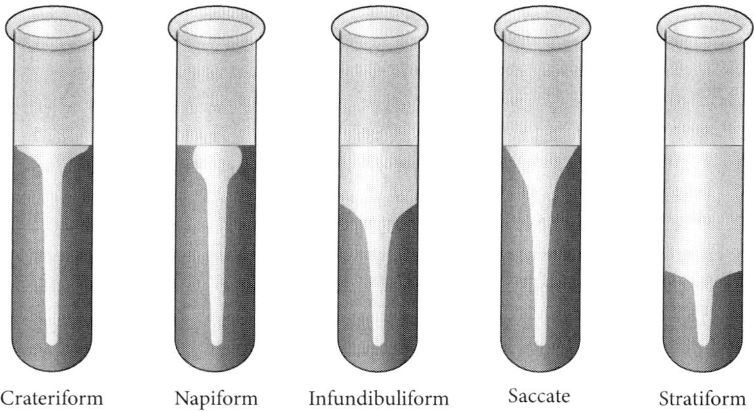

Figure 5-4 | Gelatin Liquefaction Patterns

Laboratory 6
Separating a Mixed Culture

In nature, microbes don't self-segregate into uniform populations. Rather, they are found as mixtures of many cell types. This is true in what you might consider "environmental" samples, but it is also true in clinical situations. Wound infections, especially in chronic sufferers, are commonly caused by more than one type of bacteria. Both environmental microbiology and clinical practice depend on being able to create pure cultures; therefore, methods for separating a mixed bacterial sample are foundational techniques.

In this lab exercise, you will use a standard streaking technique upon a solid medium to separate out a mixed bacterial culture. The culture is a mixture of 10% *Escherichia coli* to 90% *Staphylococcus aureus* — a 1:9 ratio. After incubation, you will observe your streak plate and use your skills in identifying cultural characteristics to assess your ability to isolate pure colonies of both species and also note which organism seems to be most abundant.

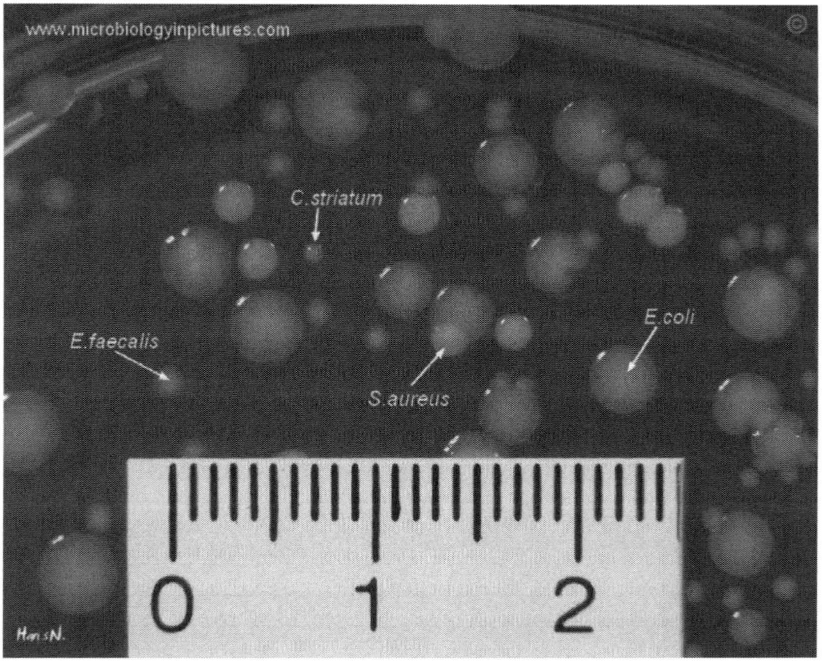

Figure 6-1 | Colonies of *E.coli*, *Staphylococcus aureus*, *Enterococcus faecalis* and *Corynebacterium striatum* on tryptic soy agar (TSA). Cultivation 24 hours, 37° C in an aerobic atmosphere.

By Hans Newman. Used with permission. https://microbiologyinpictures.com/bacteria-photos/escherichia-coli-photos/bacterial-colonies-on-agar.html

Procedure

1 | Obtain a tube of the mixed culture and two TSA plates.

Streak Plate

2 | Observing proper aseptic technique, transfer a loop of the mixed culture to the TSA plate and perform a standard streak.

Spread Plate

3 | Place the second TSA plate upon the turntable.

4 | Observing proper aseptic technique, use your pipette to transfer 100 µl of the mixed culture to the center of the plate.

5 | Remove a glass spreader from its ethanol bath and use the Bunsen burner to burn the alcohol off. ***Do not*** hold the glass in the flame; simply use the flame to light the alcohol then remove it from the flame. Leaving the spreader in the flame will overheat the spreader and can kill the culture as you spread it over the plate.

6 | Allow the spreader to cool for at least 30 seconds, then use it to spread the mixed culture over the entire surface of the plate.

7 | When finished, place the spreader back into its ethanol bath.

8 | Incubate at 37° C for 24 hours.

9 | After incubation, observe the plate.

  a | Can you discern two different colony morphologies?

  b | Note whether you managed to obtain pure colonies of either species.

  - *E. coli* is tan, irregular, undulate, and raised.

  - *S. aureus* is a lighter tan or yellow, circular, entire, and convex.

  c | Which species is more abundant? Does that comport with what you started with? Why?

  d | Which technique was better suited for separating and observing the colonies and their morphologies?

7

Laboratory 7
Microbiology of Personal Care Products

Macroorganisms, including humans, carry with them a panoply of **natural microflora.** You are carrying this collection of microbes on your skin, in your mouth, your upper respiratory tract, your urogenital tract, and your colon. What counts as "natural" varies depending on body location, between individuals, between geographical locales, and through time. This means that any object that comes into prolonged or significant contact with any part of your body or natural body openings is likely to carry away with it a sample of your microflora.

These organisms are living in some type of **symbiosis** with you. These symbioses may be one of three types: **commensal, mutualistic,** or **parasitic.** In commensal symbiosis, one party derives some benefit from the relationship, and the other derives nothing either positive or negative. In mutualism, both parties benefit from their association. In parasitism, one party benefits while the other is harmed

In the case of your microflora, the vast majority of them are commensal or mutualistic. Some common natural microflora bacterial species come from the genera *Staphylococcus* (skin), *Streptococcus* (urogenital/pharyngeal), *Neisseria* (pharyngeal), *Enterobacter* (colon), and *Proteus* (colon). There is a chance, however, that you may be carrying parasitic (pathogenic) organisms such as Group A β-hemolytic *Streptococcus, Serratia marcescens,* or Methicillin-resistant *Staphylococcus aureus* (MRSA).

In this laboratory exercise, you will inoculate a solid medium with one of your own personal care items such as a razor, deodorant stick, toothbrush, loofa, make-up brush, or something else that has prolonged contact with your body. A wristwatch or glasses will work in a pinch. We will be incubating these organisms at 37° C, average human body temperature. Under these conditions it is possible that some samples may come back positive for one or more organisms that are potentially pathogenic.

You must be very careful to observe every facet of aseptic technique when working with the plates after incubation. The colonies on the plates are comprised of many millions of cells, which is much more than what you would normally be exposed to. Thus, an organism that is not commonly pathogenic may cause an infection just by virtue of the extremely high concentration of cells which can overwhelm your immune response.

The medium you will use in this exercise is mannitol salt agar (MSA), which is a selective and differential medium. MSA is selective because it contains 7.5% NaCl which will inhibit the growth of most organisms except for halophiles such as *Staphylococcus.* It is differential because it contains the carbohydrate mannitol and phenol red, a pH indicator that will turn from red to yellow in the presence of acid. A bacterium originating from human skin, which is also halophilic and able to ferment mannitol, is likely to be *S. aureus.* Therefore, any colonies that turn the medium yellow are presumed to be *S. aureus* which is a potential human pathogen. Don't be alarmed if your sample turns up *S. aureus;* up to 30% of humans are carriers.

Procedure

1 | Obtain an MSA plate.

2 | Inoculate the plate using your personal care item.

    a | If a razor or make-up brush, apply directly to medium surface.

    b | If powder makeup, tap brush on the edge of the plate to knock some dust onto the medium surface.

    c | If something like a loofa, apply directly to surface or use a swab to sample near the center.

    d | If something like glasses or a watch, use a swab to sample an appropriate surface.

3 | Incubate at 37° C for 24 hrs.

4 | Observe after incubation and determine whether any organisms grew and whether they are likely *S. aureus.*

8

Laboratory 8
Basic Microscopy

Microscopes are technology and microscopy is a discipline. Both are an outgrowth of a very old area of study: optics. The term optics comes from the (transliterated) Greek *optikē* which means "appearance" or "look." Optics began as a study into how human sight works, but it has become much more than that over the millennia.

From the study of optics first came eyeglasses (sometime around 1289 AD), then spyglasses (first attempt at patenting was in 1608), followed by telescopes (work which also started in 1608), and finally, around 1665, it led to the subject of this chapter: microscopes. Robert Hooke (England) and Anton van Leeuwenhoek (Netherlands) were contemporaries who separately developed two types of microscope. Leeuwenhoek's instrument used a single biconvex (spherical) lens, and Hooke's used *two* convex lenses. It is in Hooke's design, the compound microscope, that all modern microscopes find their origin.

All modern microscopes are compound in design, which is to say they use more than one lens. In this laboratory, we use some of the simplest and least expensive modern scopes: brightfield microscopes (Figure 8-1). These are part of an extended family of tools, all of which have their own benefits and liabilities.

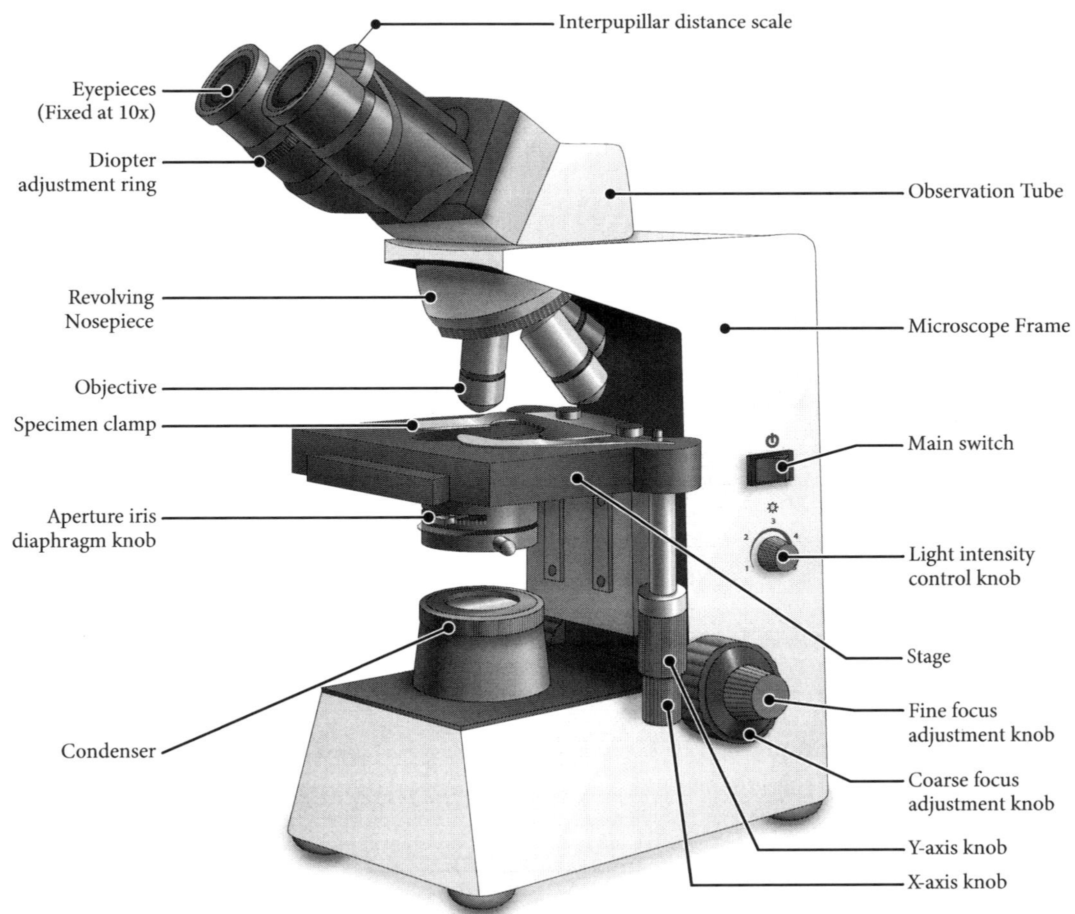

Figure 8-1 | Standard Brightfield Microscope Diagram

Modern Microscopes

Brightfield

These microscopes may have two lenses but are more commonly found with five or more, most of which are housed in the **objective.** In this type of instrument, specimens are illuminated from below via a light source that is itself focused by another lens below the stage called a **condenser.** This results in specimens that appear dark against a light background. In smaller organisms, like live, unstained bacteria, this lighting arrangement can be an issue. A brightfield system has limited contrast, which means that the differences that can be drawn between specimens and their surrounding medium are small. This is a problem when observing live organisms. Most observations of bacteria are therefore made on nonviable, stained preparations, and under oil immersion (Figure 8-2) because of their small size (Figure 8-3).

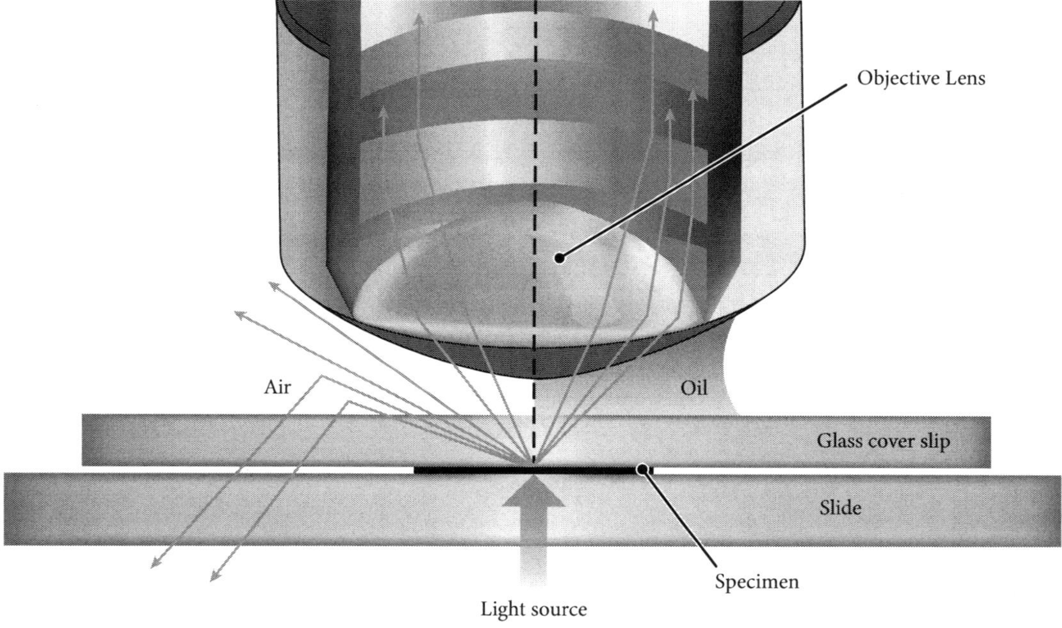

Figure 8-2 | The Effect of Immersion Oil

Darkfield

These microscopes have an ocular lens system essentially identical to brightfield microscopes. The main difference is in the design of the illumination/condenser system. In darkfield scopes, light is directed upon the specimen from the side so that light is reflected off of it rather than transmitted through it. This yields a light specimen on a dark background. These instruments are able to create greater contrast between specimens and their surrounding medium, so they are much more useful for observing both larger, opaque specimens and live microbial specimens.

Phase-Contrast

These microscopes include specialized objectives and condensers that accentuate subtle differences in the refractive indexes of cellular components. Amplifying these differences raises the contrast of various intracellular components. This is particularly suited for observing live, unstained microorganisms. These specimens appear dark on a light background.

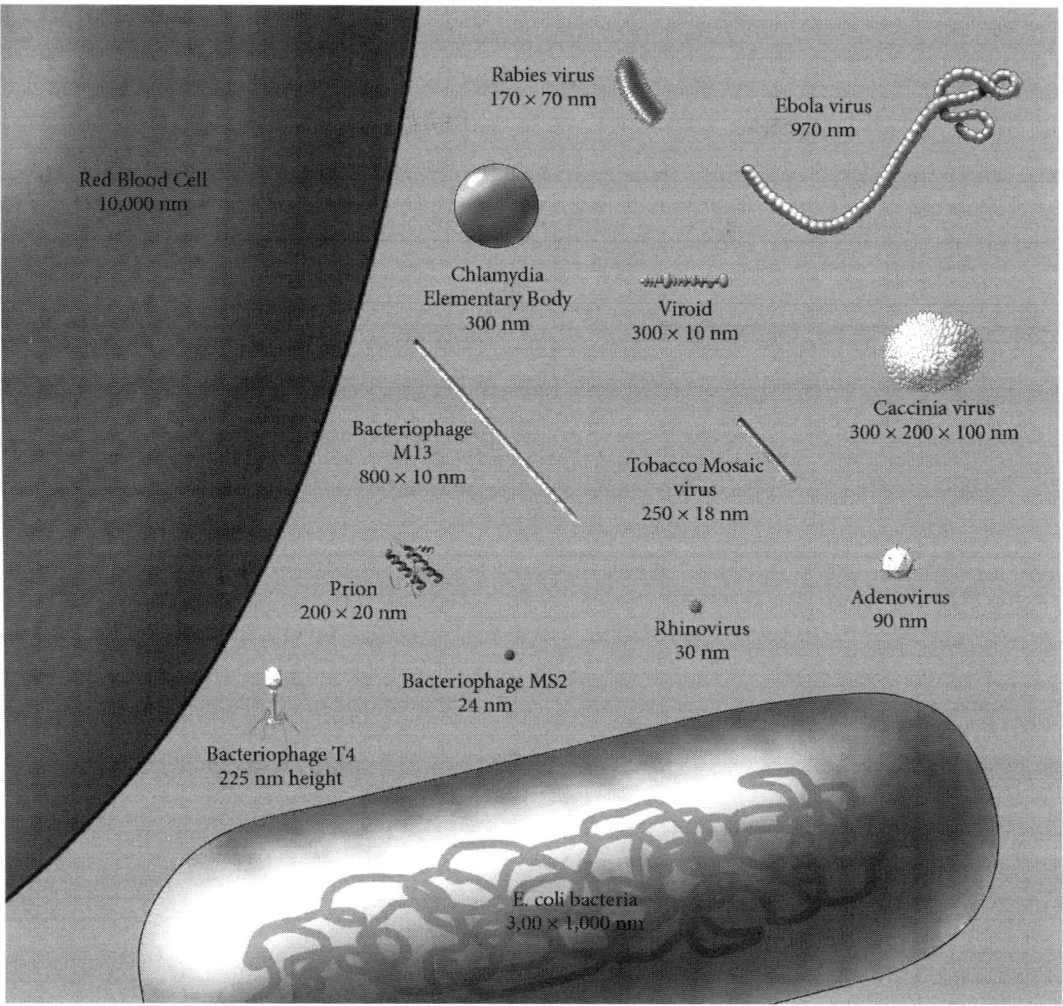

Figure 8-3 | Illustration Showing the Relative Sizes of a Red Blood Cell (Left), Bacteria (Purple, Bottom) and Various Viruses

© Monica Schroeder/Science Source

Fluorescent

These microscopes can generally function like standard brightfield microscopes, but they also have some wider uses. They have specialized light sources which are designed to cause fluorescent dyes to glow. These scopes also include optical filters that will block out the excitation light while allowing the fluorescent light to be observed, making it easier to detect your chosen response. These scopes are used in conjunction with any of a large number of fluorescent dyes wherein the microbial specimens are stained with the dye, then observed for fluorescence. This is particularly useful in observing specimens containing very small cells or contaminating objects (like soil particles) that would otherwise make delineating organism from non-living particle impossible.

Electron

These microscopes use beams of electrons and not visible light to create images of their specimens. They also do not use traditional optics, since traditional optics would absorb or scatter all of the electrons. Therefore, the optics used here are composed of electromagnetic fields, and the distance from the electron source and the specimen must be put under vacuum because even air would disturb the beam. There is more than one type of electron microscope, but they all work by differentiating between areas of different electron absorbance or reflectance.

Laboratory 9
Preparation of Microbiological Smears

Specimens of microorganisms must be prepared in some way if they are to be visualized by a microscope. The preparation procedure is determined by the organism in question, the type of microscope you will be using, and what you want to see. Such procedures might include suspending them in liquid media, using special dyes that fluoresce when struck by light of a certain wavelength, or freezing the specimen in liquid nitrogen. In this lab, the basis for most of our microscopic observations will be the **bacterial smear.**

A bacterial smear is exactly what it sounds like: a sample of a bacterial culture or colony smeared onto a microscope slide. Such preparations are decidedly low-tech, but they are the backbone of microbiological observations.

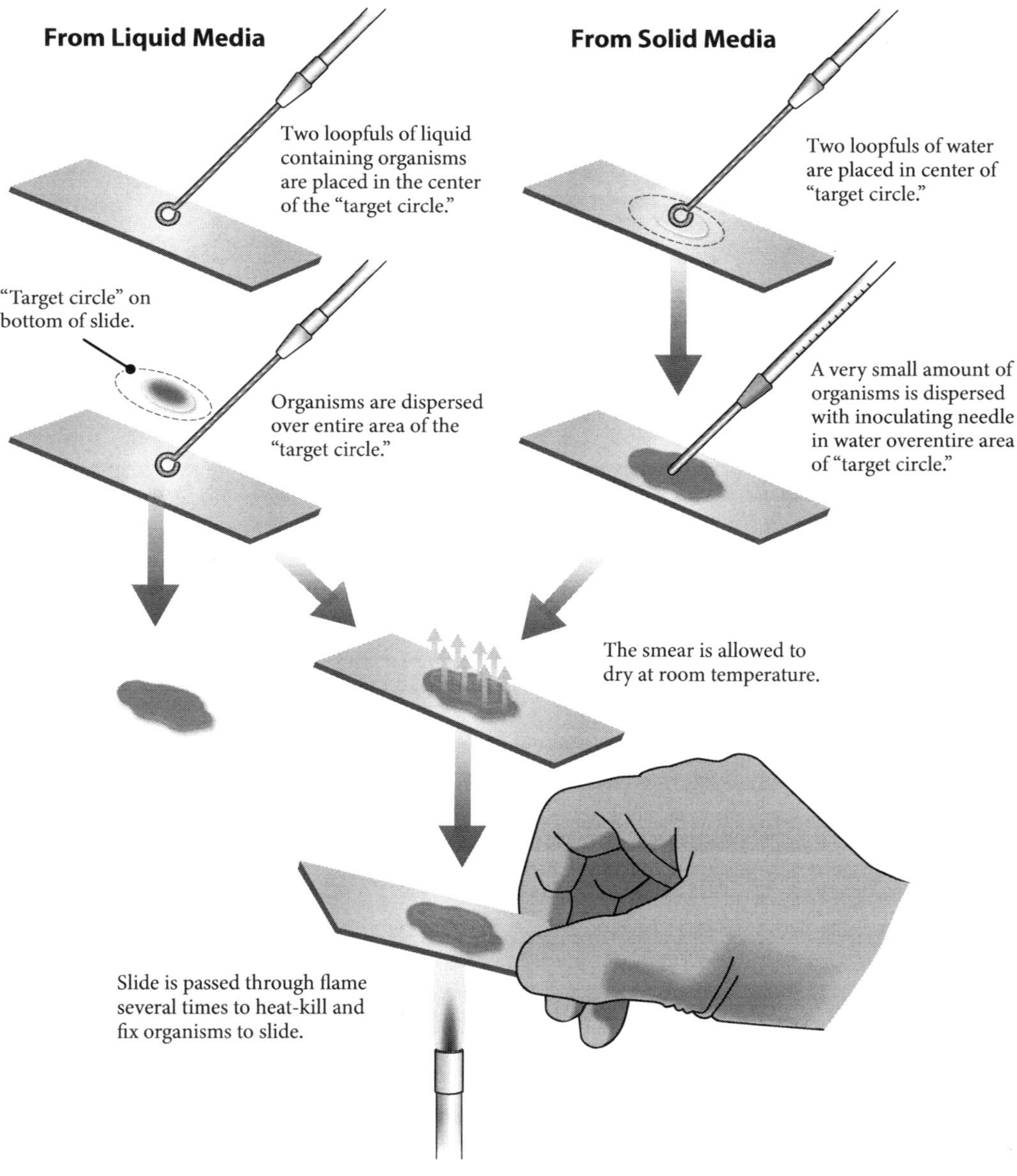

Figure 9-1 | How to Prepare and Heat Fix a Bacterial Smear

Procedure

1 | Obtain a new microscope slide, handling only the sides so as to avoid transferring oil from your fingers onto the work surface.

2 | Light up a Bunsen burner.

3 | Remove the packaging oil from the slide by passing the top and bottom surfaces briefly through the flame. If the slide has other soiling on it, you can wash the slide using soap and water but be sure to thoroughly dry the slide before proceeding.

4 | Label one of the short edges of the slide with a marker pen. Be sure your label is one that will allow you to determine whether you have the slide right side up.

5 | Obtain a bacterial culture(s).

6 | Take some amount of the culture and smear it onto the slide such that you have a smear 2–3 cm wide.

    a | If you are working with liquid cultures, use a transfer loop and transfer no fewer than 5 loopfuls of the liquid to the slide; you may need to use more than 5 depending on the organism and the concentration of cells.

    b | If you are working with colonies on agar, add 20 µl of water to the slide prior to transferring the colony sample. Use the water to dilute and homogenize your smear. Too much water will make a mess, and no water will probably yield a smear that is too concentrated to properly observe.

7 | Allow the smear to dry completely.

    a | You can let it air dry on the staining rack.

    b | If you want to speed it up a bit, you can *lightly* heat the bottom of the slide to encourage the water to evaporate quicker. If you heat it too much, you will cause the bacterial cells to lyse, and you will therefore see nothing when it comes time to observe them.

    c | ***Do not blow*** on the slide or wave it around. This encourages contamination of the sample.

8 | The smear should be semitransparent. If it is translucent or opaque, it is too thick, and you'll need to make another.

9 | Once the smear is completely dry, **heat fix** the smear to the slide by heating the underside of the slide in the flame. It should get hot enough that you could burn yourself if you touched the wrong end but not so hot that you crack the slide or char the smear.

10 | Set smear aside to cool before moving onto the next exercise.

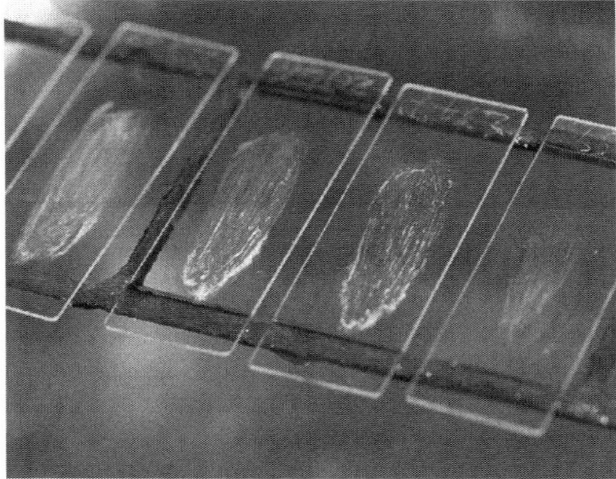

Figure 9-2 | Well-Prepared Bacterial Smears

© Andy Crump, TDR, World Health Organization/Science Source

Laboratory 10
Principles of Staining Microorganisms

Because bacteria are so small and have such low contrast, bacterial smears are generally useless unless stained. The specific staining procedure that ought to be used is determined by the type of organism present in the smear and the types of questions you want to answer about the objective.

"Stains" are actually dyes. The ones that are commonly used in microbiology are usually based on a benzene ring and are composed of three basic parts:

- **Benzene ring** – Colorless but very versatile scaffold to which you may attach functional groups.

- **Chromophore** – The portion that imparts the color.

- **Auxochrome** – The charged portion that determines the binding properties. If this is omitted from the compound, you will have a chromogen (colored compound) and not a stain.

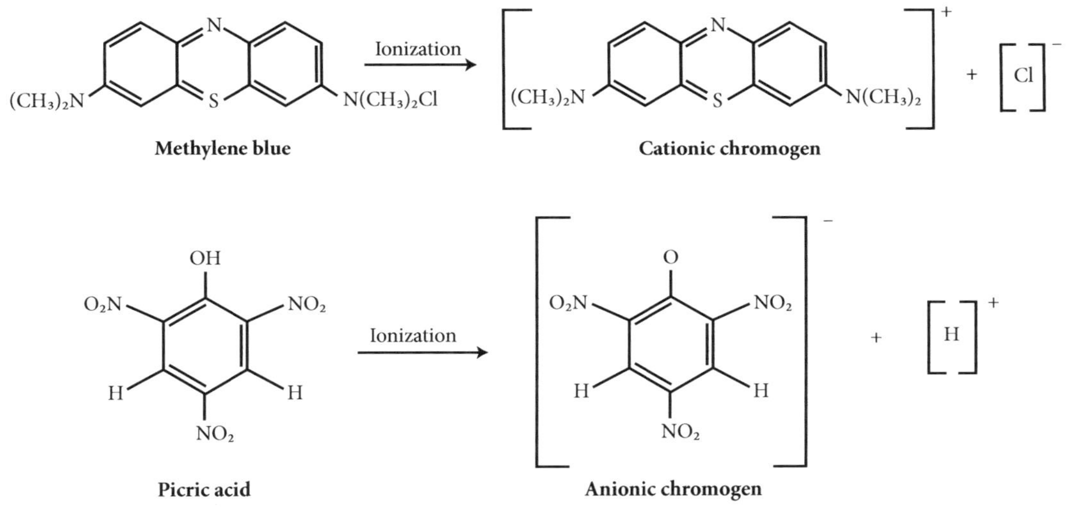

Figure 10-1 | Bacterial Staining. Examples of cationic (basic) and anionic (acidic) stains.

In this lab we use two major types of stains:

- **Acidic** – Stains that carry a negative charge (anionic). This includes compounds like eosin, picric acid, acid fuchsin, India ink, and nigrosin. These stains have the same charge potential as the cell walls and membranes of microorganisms. This means that the stain is generally not taken into the cell. Stains like eosin stay outside the cell where metabolites like acids can interact with them. Others, like India Ink, are used for negative staining techniques like capsule staining.

- **Basic** – These are positively-charged (cationic) stains that have a strong affinity for negatively-charged cellular components like bacterial cell walls and nucleic acids. Stains like crystal (gentian) violet and methylene blue are basic stains that will easily and quickly stain bacterial cell walls for visualization. Others, like safranin, are used as counterstains in differential staining procedures like the Gram stain. Basic fuchsin is used in acid-fast staining.

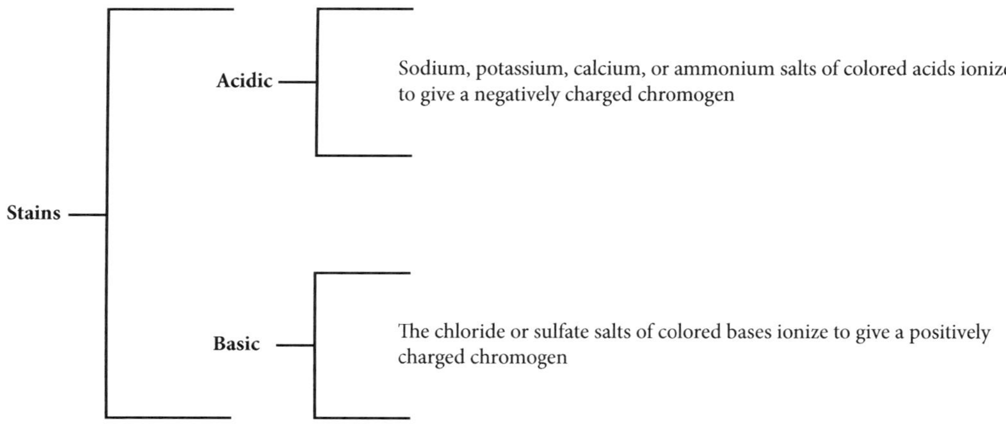

Figure 10-2 | Acidic and Basic Stains by Type of Salt

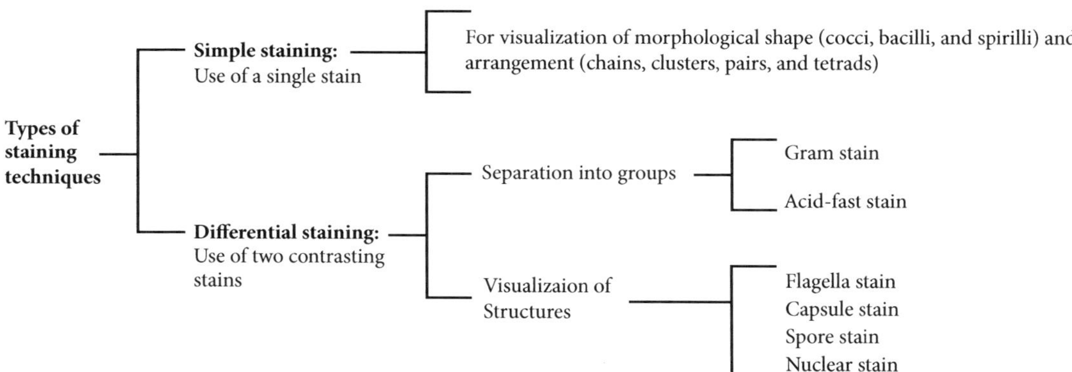

Figure 10-3 | Staining Techniques

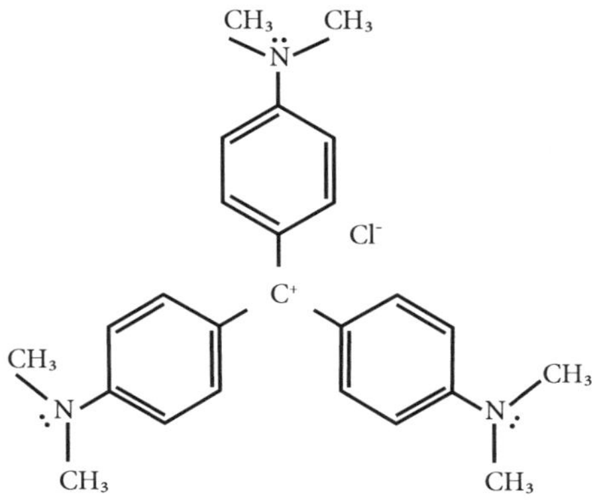

Figure 10-4 | Crystal Violet

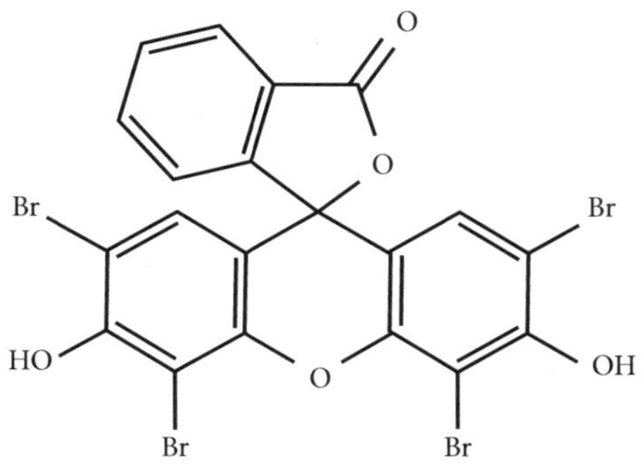

Figure 10-5 | Eosin Y

These stains are used in many types of staining procedures, but the only two that we will use in this lab are **simple staining** and **differential staining.** Simple staining typically includes the use of one stain and is useful for observing cell **morphology** and **arrangement.** Differential staining is used to allow you to differentiate different types of bacteria, so they can be classified into different groups based on their biochemical anatomy.

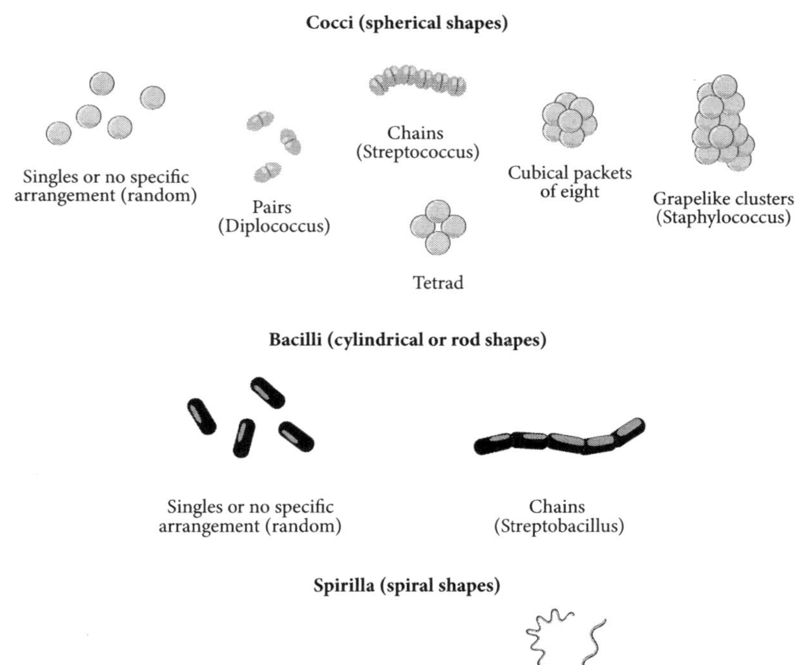

Figure 10-6 | Bacterial Morphology Shapes

Laboratory 11
Simple Staining with Basic Dyes

Stains increase the contrast of microbial cells which allows us to more accurately and easily observe their gross properties. For the purpose of counting or identifying the morphology and arrangement of a particular bacterium, a simple staining procedure is adequate.

Procedure

1 | Obtain your prepared smear(s).

2 | Place your smear on the staining rack smear side up.

3 | Flood the surface of the smear with crystal violet or methylene blue and let it stand for 30 seconds to a minute (note how much time you give it).

4 | Carefully rinse the excess stain off into the staining tray using a water rinse bottle.

5 | Allow the slide to air dry (or help it along with fire).

6 | Place on the microscope stage and add a drop of emersion oil directly to the smear.

7 | Use the 100x objective and move the stage up until oil flows onto the objective lens.

8 | Turn your light all the way up, your iris/diaphragm 60% closed, and look through the eyepieces while you make quick, clockwise, quarter turns of the fine focus knob.

9 | Be careful that you don't overshoot the focal plane of your smear. If you go too far, you can cause the objective to come into contact with the slide. The objective lens is spring-loaded, so it can take a little abuse, but if you press too far, you will crack your slide and may scratch the lens coating which will ruin it.

10 | Observe and report in your lab notebook the morphologies and arrangements of all the bacterial cultures being used in the lab.

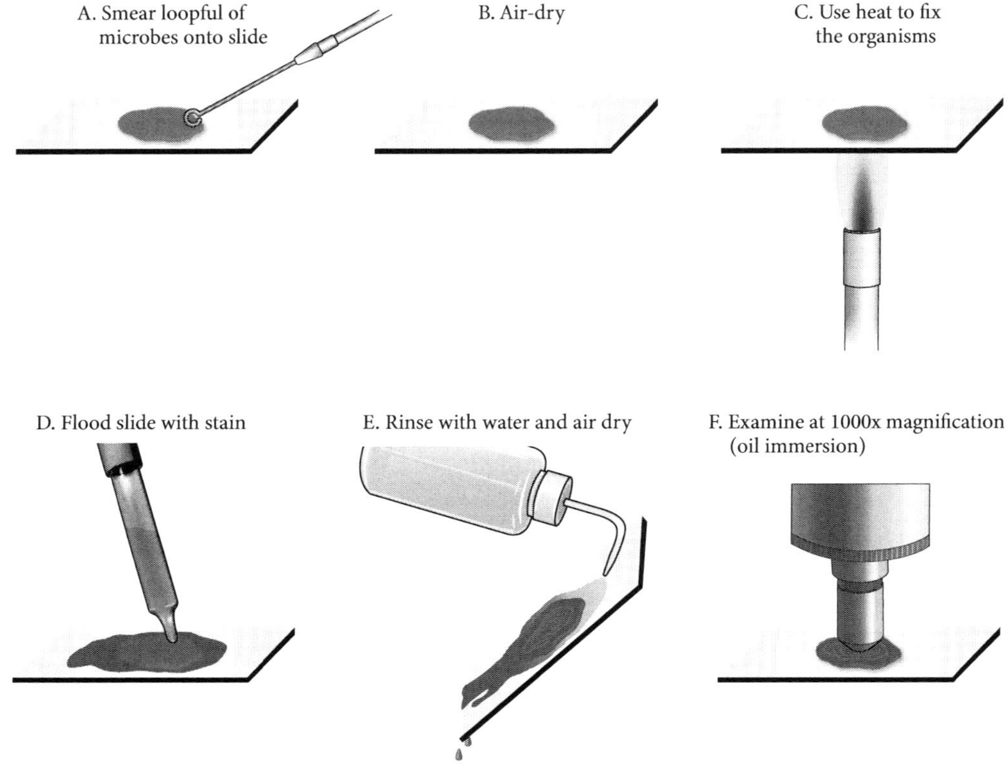

Figure 11-1 | Simple Staining

12

Laboratory 12
The Gram Stain: A Differential Dichotomy

The identification of microorganisms has been a major undertaking since microorganisms were first shown to exist. Microbes can be split into categories based on a number of physiological, biochemical, or anatomical properties, and these categories can have important clinical significance. In the case of bacteria, currently, the final word in **taxonomic identification** is DNA sequences, but unless you are willing to pay well above the going rate, it takes no less than two days to generate the data. In clinical applications where time, cost, and throughput can be critically important factors, physiological and morphological differences are the big players.

One of the most important steps in identifying a clinically-significant bacterium is to determine if it is **gram negative** or **gram positive.** The quickest and most cost-effective way to determine this is via the **Gram stain.** The Gram stain is a **differential staining** technique that can tell us about the arrangement of bacterial cell walls. When properly executed, the Gram stain will yield gram-positive (gram+) cells stained purple and gram-negative (gram–) cells stained red (sometimes can appear pink/orange if the microscope light isn't perfectly white). If performed on acid-fast bacteria, the Gram stain will cause the cells to come out both ways, which is called **variable staining.**

This procedure is named for its inventor, Dr. Hans Christian Gram, though it has undergone revision and refining since he published the first protocol. The basics have remained the same, however:

- **Primary stain** – A basic stain that has affinity for the **peptidoglycan** of bacterial cell walls. We will use crystal violet.

- **Mordant** – A compound that will combine with the primary stain to form a complex that is water insoluble. We will be using Gram's iodine.

- **Decolorizer** – A solvent that will wash away any primary stain and stain/mordant complex that is not bound up in peptidoglycan. We will be using 95% ethanol (EtOH).

- **Counterstain** – Another basic stain with a coloration distinct from the primary stain. This will re-stain any cells that were made colorless by the decolorizer, but the color will be different from the primary stain. We will be using safranin.

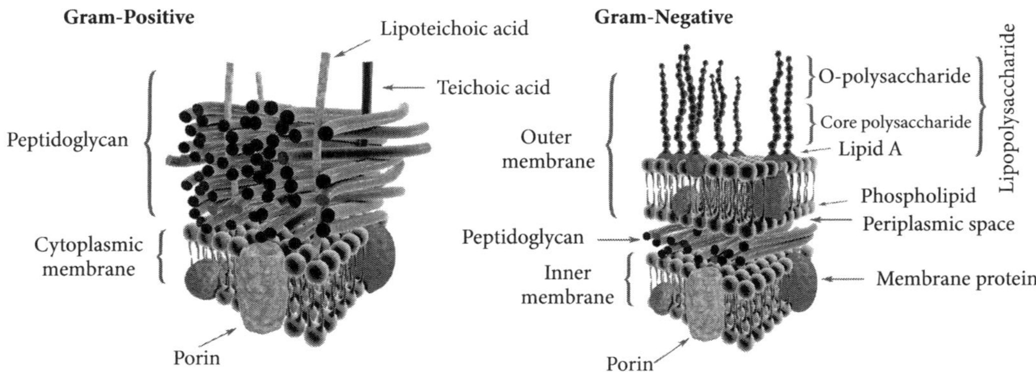

Figure 12-1 | Gram-Positive vs Gram-Negative Cell Walls

The staining differences between gram+ and gram– are primarily a function of the thickness of their cell wall and whether they have an outer membrane. Gram+ cells have much thicker cell walls which consist of many overlapping layers of peptidoglycan cross-linkages which trap a significant amount of primary stain/mordant complex leaving them stained purple after decolorizing. Gram– organisms have a thinner cell wall, and it is buried below an outer membrane that acts as a gatekeeper that excludes primary stain. This means that virtually all the primary stain/mordant complex will be rinsed away by the decolorizor, leaving the cells colorless and ready to accept the counterstain.

Procedure

You can proceed through these steps without waiting for the slide to cool or dry.

1 | Create a bacterial smear. Be sure to burn off the packing oil and firmly heat-fix the smear because the ethyl alcohol (EtOH) will weaken the bond of the smear to the glass.

2 | Place smear on the staining rack and flood the smear with crystal violet; let stand for 1 minute.

3 | Rinse excess stain off with a rinse water bottle.

4 | Flood with Gram's iodine and let stand for 1 minute.

5 | Rinse mordant off with water.

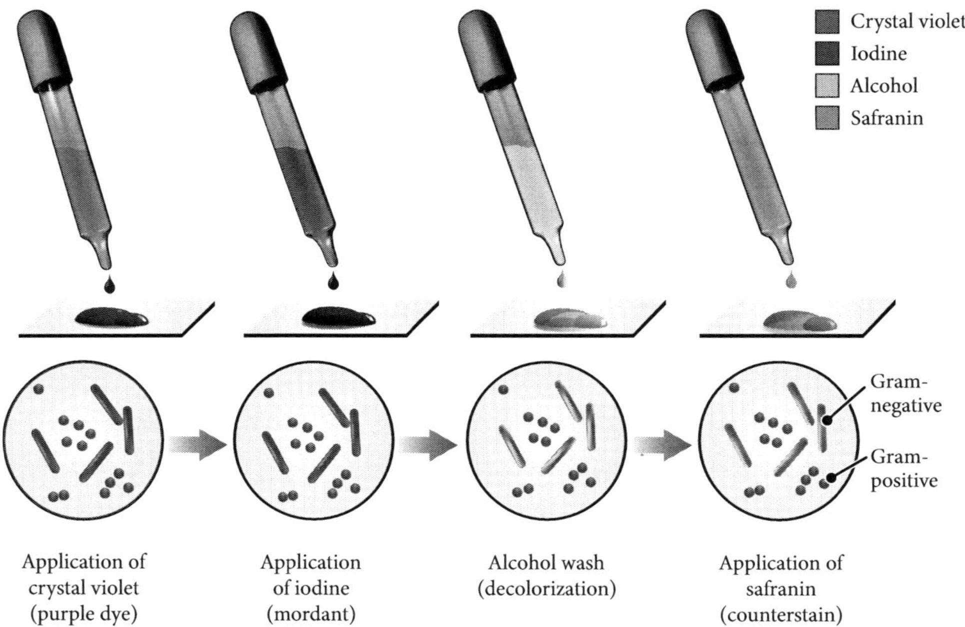

Figure 12-2 | Gram Staining

6 | Decolorize the smear by introducing EtOH drop-by-drop until the drops that fall off the slide are 95% clear of purple.

7 | *Gently* rinse the EtOH off the slide using water.

8 | Flood with counterstain and let stand for 1 minute.

9 | Rinse the counterstain off with water.

10 | Dry the slide and observe under 1000× oil immersion.

Finally, beware of performing the Gram stain on cultures that have been exposed to antibiotics such as penicillin or are older than 24 hours. These sorts of conditions can cause bacterial cells to alter the construction of their cell walls causing gram+ to appear gram– or gram-variable.

13

Laboratory 13
Wet Mounts

Creating and staining bacterial smears causes the organisms to become affixed and stationary. They also become dead. Indeed, that's the main point of the procedure because we want them to stay put! There are, however, important aspects of microbes that can only be observed when the organisms are alive and free to move. To observe these properties, we will create **wet mounts.**

Wet mounts are exactly what they sound like: a liquid suspension of microbes on a slide. Bacterial cells are small, and their refractive index is close to that of water. These facts combine to make it challenging to observe them when wet mounted. Rise to that challenge, and you'll be rewarded with observations of **motility** (if the organisms have **flagella**), **binary fission, Brownian Motion** (the vibration of particles and cells as they are struck by water molecules), and the natural size and shape of the cells absent heat and chemical treatments.

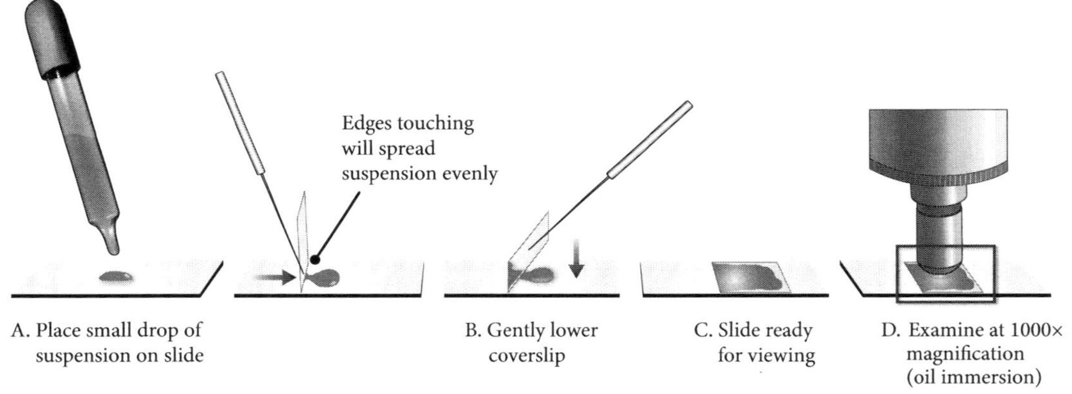

Figure 13-1 | Creating a Wet Mount

Procedure

1 | Obtain a slide and burn off the packing oil.

2 | Add 20–25 µl of a liquid bacterial culture to the slide.

3 | Place a coverslip over the liquid.

4 | Observe at 1000x under oil immersion.

Some Final Considerations

Do one wet mount at a time because they dry out if you let them sit too long. Also, you'll be observing your wet mount using a standard brightfield microscope, but we would find it easier, though significantly more expensive, to make these observations if we were able to use phase-contrast microscopy.

Laboratory 14
Serial Dilution Plate Counts

It is often important to determine a quantitative estimate of how many microbes are present in a given environment. This is an especially important measurement to make during the manufacture of things like cosmetics and food or during key processes like wastewater treatment.

There are a lot of ways to estimate microbial abundance, and they all have their own strengths and weaknesses. We'll discuss just 6 different methods here to give a little context.

- **Direct microscopic counts** – Making direct counts of bacterial cells requires either a specialized slide called a **Petroff–Hausser counting chamber (hemocytometer)** (Figure 14-1) or an eyepiece lens that includes a grid. The PH counting chamber is a thick glass slide designed to suspend a known amount of liquid sample over a counting grid. An eyepiece with a grid can be used to enumerate liquid, fixed, or filtered samples, but the operator must calibrate the grid area at different magnifications using a stage micrometer. These methods of enumeration have the advantage of being quick, but they are not inherently sensitive to whether the cells are alive or dead.

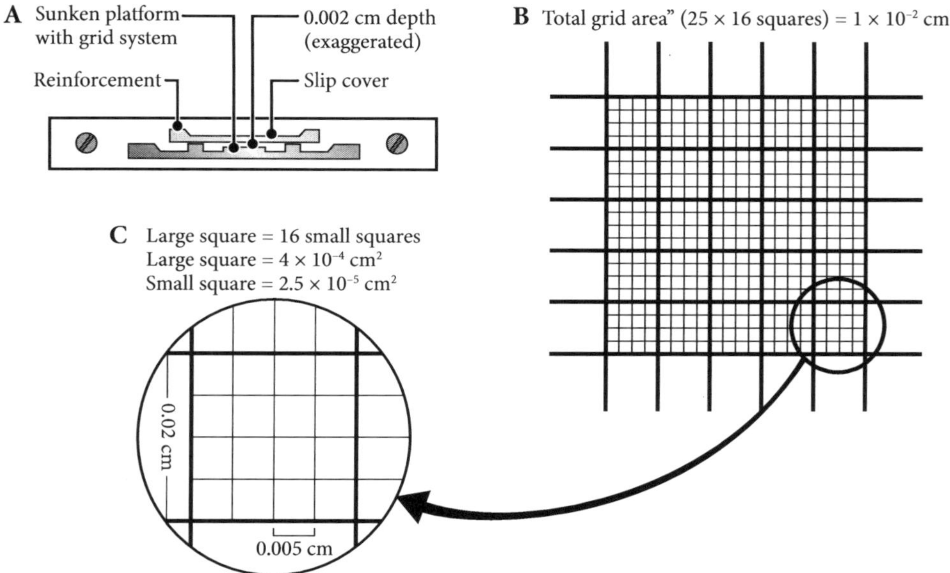

Figure 14-1 | Petroff–Hausser Counting Chamber

- **Coulter counter** – This is a bit of specialized equipment that can infer the number of cells in an electrolyte. It does this by passing fluid containing cells (or any particle) through a small aperture through which is flowing an electrical current (Figure 14-2). Cells are non-conductors so as they pass through the charged aperture, they increase the electrical resistance which is being recorded. This method has the benefit of being rapid and of not requiring the investigator to spend time at a scope making counts. Major disadvantages are that the machine can't distinguish live cells from dead cells and neither can it definitively tell you what is a cell and what is non-living particulate matter.

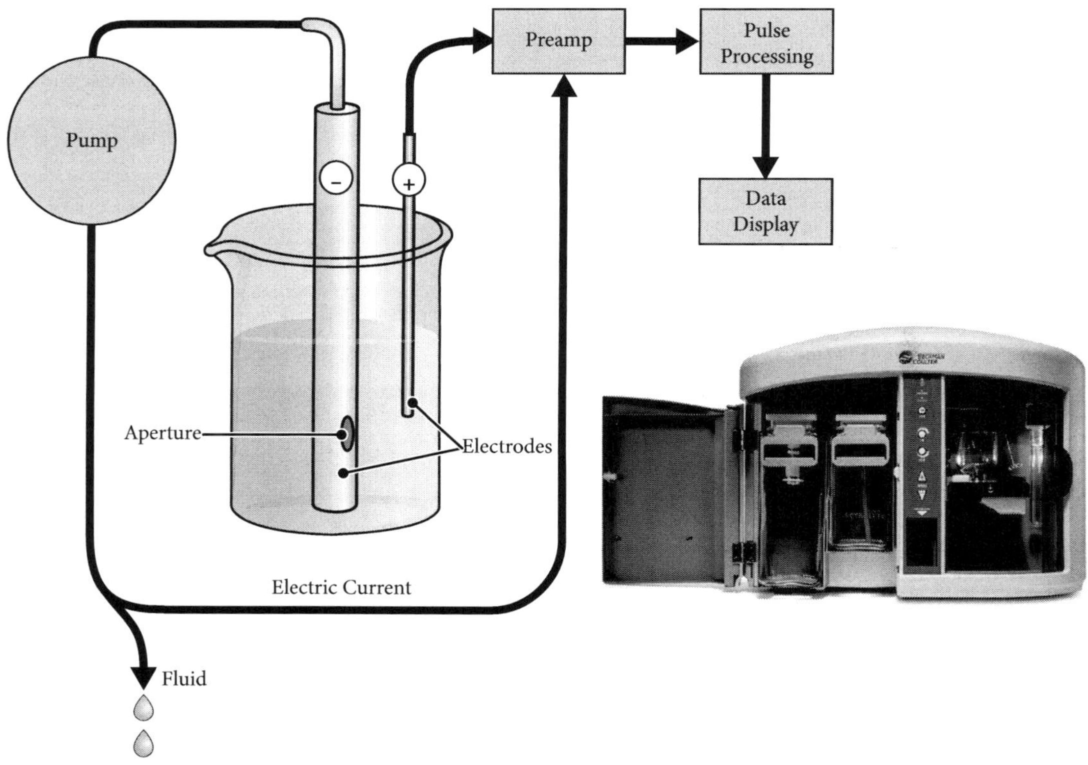

Figure 14-2 | Coulter Counter

Right: Used with permission from Beckman Coulter Life Sciences

- **Flow cytometer** – This is another piece of specialized equipment that detects cells as they pass through a narrow channel (Figure 14-3). A flow cytometer has a principle similar to that of a Coulter counter except that it uses lasers to detect the particles flowing through the detection channel. Flow cytometry has the benefit of using optical technology which can detect the size of each particle and can also be used to detect stains or natural pigments, and this can help cut out noise from non-cellular particulate matter. Advanced versions of these machines can also sort the cells based on their optical characteristics.

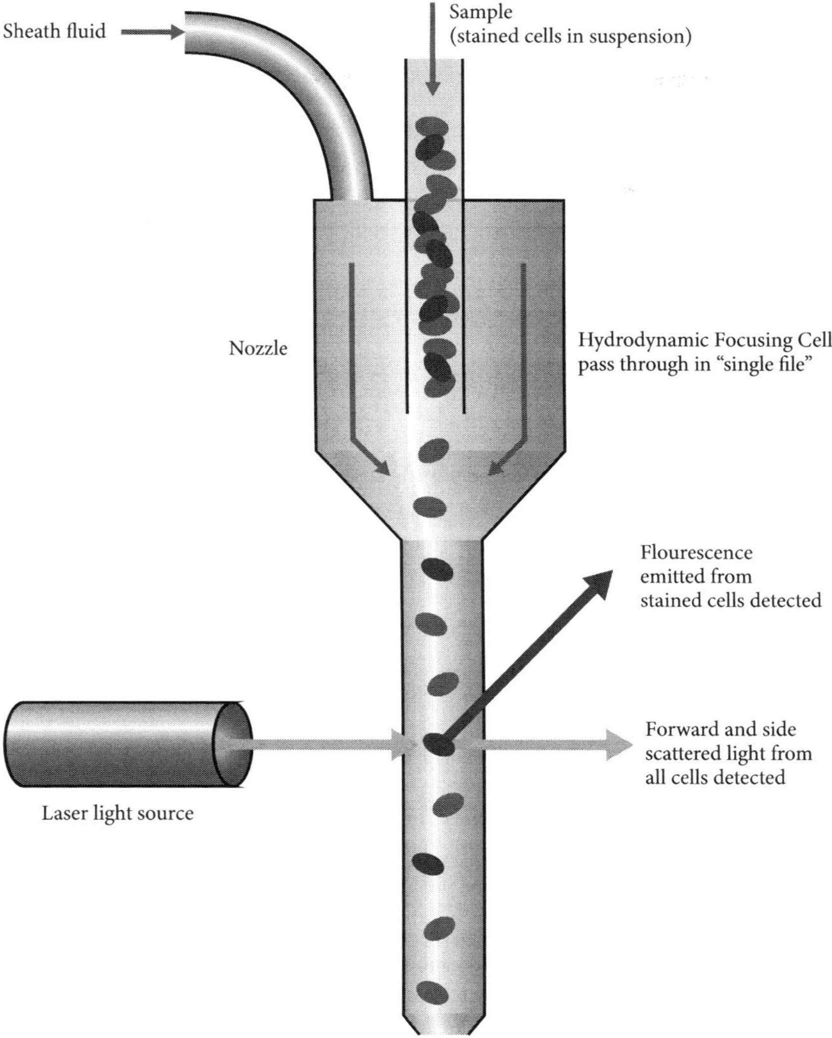

Figure 14-3 | Flow Cytometry

- **Spectrophotometric analysis** – This is another optical method wherein the **turbidity** (or **optical density**) of a microbial culture is used as an analog for cell number (Figure 14-4). Cells absorb and deflect light so that, as their population grows, the turbidity (cloudiness) of the culture increases, and this can be quantified by a spectrophotometer. These measurements are very quick, non-invasive to the sample, and inexpensive. The major disadvantages are that it is a relatively insensitive technique which means you need suspensions of 10 million cells or more to get detectable readings, and you aren't directly counting cells so you have to infer cell number. To be fair, however, turbidity can be linked to cell number if direct counts are performed concurrently with the optical measurements.

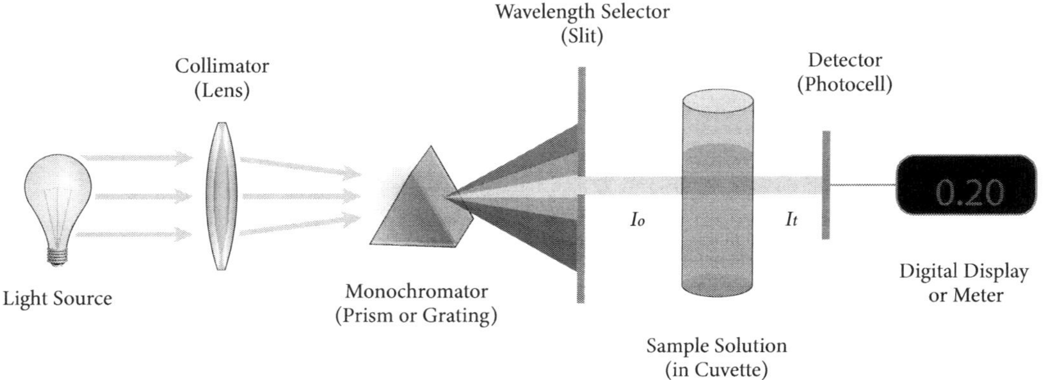

Figure 14-4 | Spectrophotometric Analysis

- **Chemical methods** – These shouldn't be considered to be quantitative measurements since they aren't always directly linked to the number of cells present. These methods include measurements of protein, fat, and DNA concentrations (spectrophotometry again) or dry weight, oxygen consumption (special electrodes or wet chemistry) (Figure 14-5), or gas (like H_2) production (gas chromatograph).

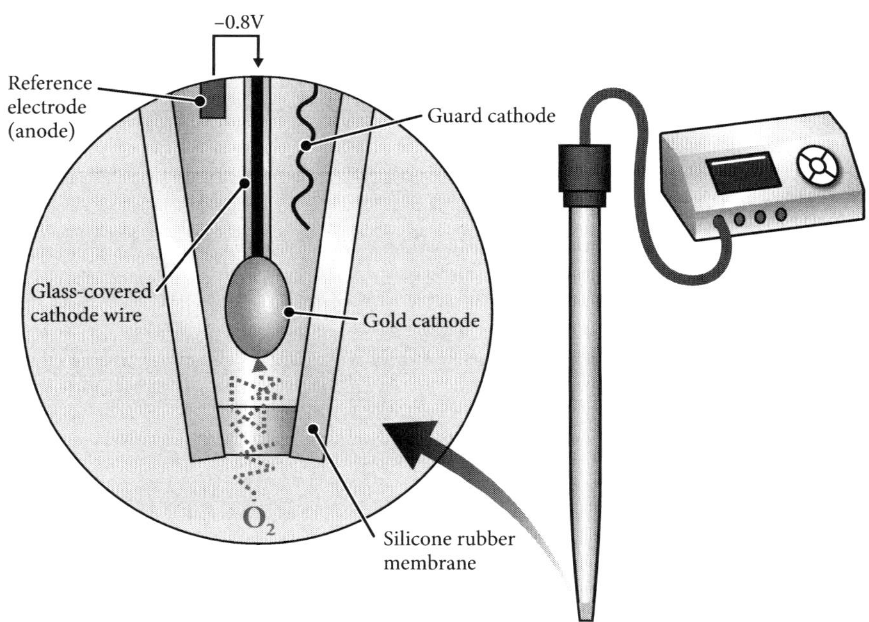

Figure 14-5 | Oxygen Consumption Electrode

- **Serial dilution plate counts** – This is the method that you will be employing this week (Figure 14-6). This is the only method listed here which can provide an estimate of the concentration of **viable cells**. In many medical and sanitary applications, it is the microbes that are metabolically active and capable of reproducing (viable) that matter. It's time consuming and requires more disposable materials and media than the other methods, but it can also yield discrete colonies that can be used to create pure cultures.

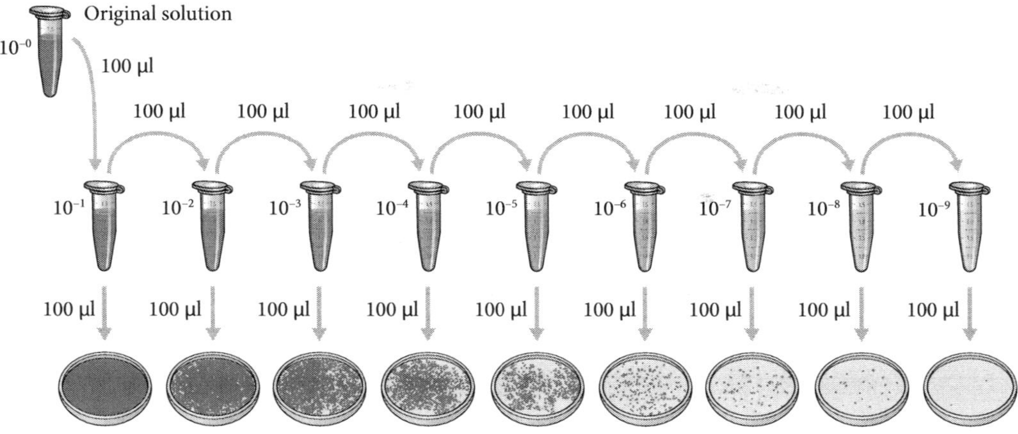

Calculation: CFU/ml = (# of colonies on plate)/[(pipetted volume in mL) $\times$ 10^{-x}]
Where x = The exponent of the dilution whose plate you are counting.

Figure 14-6 | Serial Dilution

Procedure

1 | Obtain a tube of liquid bacterial culture, enough tubes of sterile saline to do all the dilution steps, a pipette, some pipette tips, and a stack of TSA plates sufficient to plate the required dilutions.

2 | Note the dilution factor of your given bacterial culture. Label it 10^0 if it is undiluted. Label the rest of the tubes in descending exponents from what you labeled the first tube, i.e. 10^{-1}, 10^{-2}, 10^{-3}, **all the way through to 10^{-9}.**

3 | Vortex the bacterial culture to re-suspend any cells that settled out.

4 | Open the bacterial culture, extract 100 µl, and transfer it to the first (e.g. 10^{-1}) sterile saline tube.

5 | Close both tubes and vortex the 10^{-1} tube.

6 | Re-open the 10^{-1} tube and transfer 100 µl to the 10^{-2} tube.

7 | Close the tubes, vortex the 10^{-2} tube, and proceed through the rest of the dilutions.

8 | Once all dilutions have been completed, **open each tube in turn** and transfer 100 µl into a single TSA plate, labelling the plates with the dilution factors as you go.

9 | Discard all dilutions in biohazard containers and incubate the plates at 37° C for 24 hours.

10 | After incubation, choose *a single plate* to count, being sure the plate has a number of colony forming units (CFU) within 30–300.

11 | Convert your CFU count to CFU/mL in the original culture according to the following equation:

CFU/ml = (# of colonies on plate)/[(pipetted volume in mL) $\times$ 10^{-x}]

Where x = The exponent of the dilution whose plate you are counting.

For example, you do your dilution according to the protocol given in this exercise. Do plating and incubate as described. After incubation, you find that that 10^{-4} plate has an appropriate number of colonies in it for counting. You count and find 222 CFU. The math would be as follows:

CFU/ml = 222/(0.0001 $\times$ 0.100) = 22,200,000

15

Laboratory 15
Culture Maintenance

A major focus of this laboratory course is to educate future healthcare and pharmaceutical professionals about how to cultivate desirable microorganisms and reduce or eliminate harmful ones. It is therefore important for you to know something about what microorganisms require to live.

The microbes you've been working with are living things, and that means that they need nutrients, energy, and water to continue living. Microbes are very diverse, so not all microorganisms have the same requirements. The parameters that some microorganisms require to thrive are bad enough for other organisms that they will die if exposed to them. It is therefore important that we know the general requirements for an average organism *and* the specific requirements for whatever is our specific organism.

In general, microorganisms require the following:

- **Water** – Every living thing we know of requires some amount of water to live and reproduce. The cytoplasm within cells is approximately 80% water and 20% dissolved substances. Microorganisms also need water outside the cell to facilitate the movement of low molecular weight nutrients across the cell membrane.

- **Carbon** – Considered a **macronutrient,** it is the element most commonly found in living things and is half of the definition of what constitutes **organic compounds.** The dry weight of living things, micro or macro, is predominantly carbon. It is important to note how organisms acquire their carbon:

  - **Autotroph** – *Auto* = "self, same, or spontaneous" and *trophic* = "pertaining to food." Organisms capable of autotrophy obtain their carbon from inorganic carbon in their environment (usually carbon dioxide) and **fix** it into organic compounds for use in their cells. These organisms are not human pathogens, except in the case of *Clostridium difficile,* which has some autotrophic genes and has been shown capable of growth with $CO_2 + H_2$ as a sole carbon source.

  - **Heterotroph** – *Hetero* = "different or other." Organisms capable of heterotrophy obtain their carbon via organic carbon (previously fixed from CO_2) in their environment. All human pathogens (with the exception of *Clostridium difficile*) fall within this class. Such pathogens obtain their carbon from *you.*

  - **Mixotroph** – *Mix* = "mingling". Mixotrophic organisms are capable of both autotrophy and heterotrophy and will engage either of them depending on environmental conditions. Many bacterial and some microalgal species are capable of mixtotrophy. Even some higher plants (macrophytes) such as venus flytraps and perhaps even animals such as the oriental hornets exhibit a level of mixtotrophy.

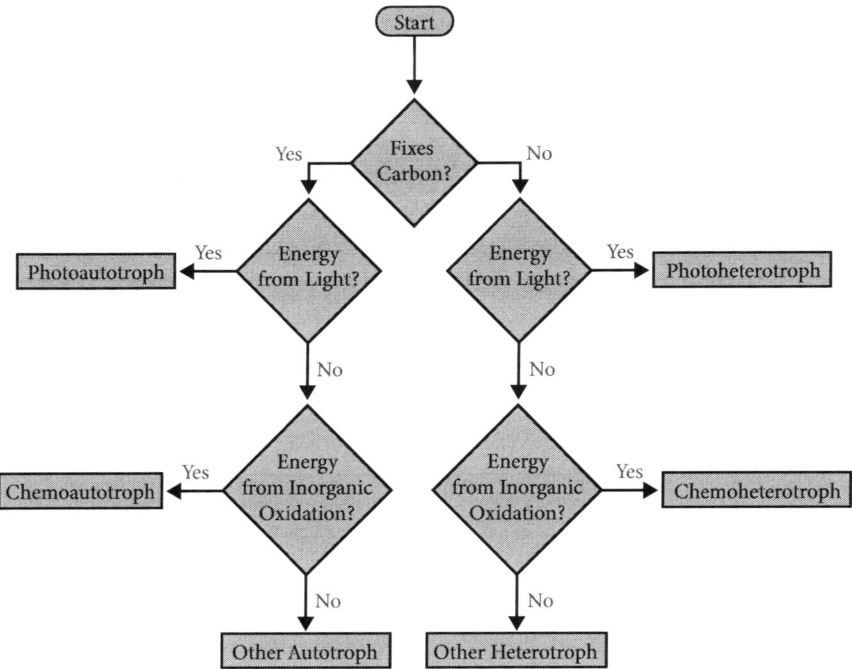

Figure 15-1 | Carbon/Organism Types

Modified from Pekaje [CC BY-SA 3.0 (http://creativecommons.org/licenses/by-sa/3.0/)] https://commons.wikimedia.org/wiki/File:Troph_flowchart.svg

- **Nitrogen** – This is another macronutrient that, in terms of mass, is perhaps the second most abundant element in microorgansms. Nitrogen is used in many essential biological compounds, especially amino acids (proteins) and nucleic acids. Proteins are the workhorses of the cell, performing essential duties, especially as enzymes that lower the **activation energy** of key chemical reactions. Nucleic acids are the medium of the genome wherein is stored all of the information necessary for the life of the given organism (DNA). They also comprise the templates for protein translation (mRNA) and are key components of the ribosomes that do the translation (rRNA), combining the amino acids into a peptide chain (tRNA). Nucleic acids can also be used like **adenosine triphosphate** (ATP) as an energy currency within the cell. There are **nitrogen fixers** capable of splitting the triple covalent bond in N_2 and making NH_3, **nitrifyers** who reduce oxidized nitrogen species, and **denitifyers** who turn bioavailable nitrogen back into N_2 gas. The nitrogen species you need to supply depends on the organism in culture: N_2 gas, NH_3, or NO_2^-, NO_3^- salts.

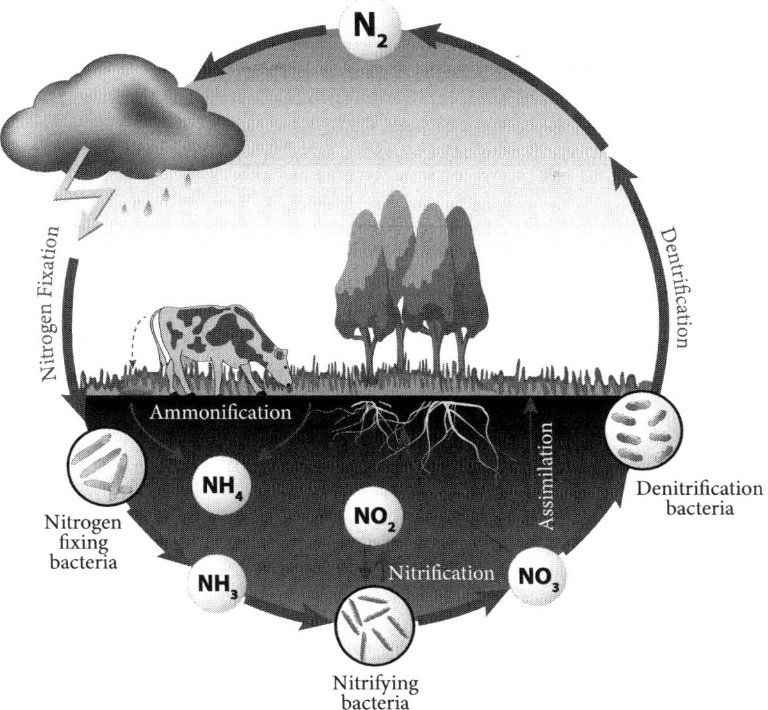

Figure 15-2 | The Nitrogen Cycle

- **Non-metallic elements** – These are mainly P and S, which are also macronutrients. Phosphorus is a key player in the energy economy of the cell and is an essential part of genomic structure. S is a common constituent of proteins, a component of some lipids, and is important in nitrogen metabolism. These are supplied to microbes as P salts, amino acids, sulfate, or elemental sulfur.

- **Metals** – These include elements like Na^+, K^+, Mg^{2+}, Ca^{2+}, Mn^{2+}, Fe^{2+}. Fe^{3+}, Cu^{2+}, Zn^{2+}. Several of these function as essential cofactors in the operation of enzymes. Others function as osmoregulators or in the transport of electrons during metabolism. Most metals are considered **micronutrients** and are usually supplied via metal salts.

- **Vitamins** – These are organic substances, usually supplied in minute amounts, that function as metabolic coenzymes facilitating efficient metabolism. The larger the profile of vitamins an organism requires be supplied in the medium, the less flexible and more **fastidious** it is. Many human pathogens are fairly fastidious.

- **Energy** – Most activities in the cell require energy input to proceed. Active membrane transport, amino acid synthesis, DNA transcription, protein translation, and motility all require a large amount of energy. Organisms can be classed according to how they obtain their energy:

  - **Phototrophs** – These obtain their energy from solar radiation. These organisms use systems of pigments that capture energy from certain photons and use that energy to excite electrons from which the energy can be harvested. These organisms will therefore require light in order to grow.

  - **Chemotrophs** – These organisms obtain their energy by oxidizing chemical compounds, either organic or inorganic. The most common organic energy source is dextrose (glucose), but there are many microorganisms that can use inorganic compounds such as H_2S or $NaNO_2$.

- **Gaseous atmosphere** – It is common for an organism to require atmospheric oxygen (**obligate aerobe, microaerophile**), but it is also common for organisms to thrive without it (**obligate/ facultative anaerobes, aerotolerant**). Other organisms, such as many human pathogens, require some specific fraction of their gaseous environment to be CO_2 up to perhaps 3%–5%. For some organisms, it is a matter of metabolic terminal electron acceptors and for others, like photoautotrophic microalgae, it's a matter of obtaining carbon for fixation. For obligate anaerobic organisms, it is a matter of excluding the atmospheric oxygen that leads to the formation of deadly peroxide and superoxide.

- **Temperature** – Every organism has an **optimal temperature** at which its compliment of enzymes runs most efficiently and allows the organism to grow at its highest theoretical rate. Every organism also has a **temperature range** over which it can grow at some rate less than the optimum. Terms like "low" and "high" are relative with regard to temperature. There are microorganisms capable of thriving in the superheated seawater deep in oceananic volcanic vents, and there are microorganisms that thrive between layers of Antarctic ice. What is a high temperature for one organism is low for another and vice versa. Low temperatures slow enzymatic reactions, slowing metabolism and growth. High temperatures can **denature** enzymes, rendering them useless and killing the organism.

- **pH** – Every microorganism has some **optimal pH** which will allow maximal growth rate and a **pH range** over which growth can occur but at a rate slower than the optimum. pH (not Ph or PH) is the negative log of the concentration of hydrogen ion ($-\log [H^+]$) in a solution. The pH of the environment can greatly impact the speed of enzymatic reactions and the generation of ATP, which proceeds via **chemiosmosis.** Like with temperature, discussion of "high" and "low" pH values is discussion of relative amounts. pH < 7 = acidic, pH = 7 = neutral, pH > 7 = alkaline (basic). A pH too far outside the range for a given organism can cause denaturation of enzymes and death.

Culture Maintenance

The maintenance of microbial cultures means managing their environment so as to keep them living and reproducing. In practice, this generally means the creation of fresh growth media, subculturing organisms in/onto it at regular intervals, and incubating the subcultures at an appropriate temperature.

These subcultures are required because as microorganisms grow on/in their medium, they use up nutrients and they export metabolic wastes that will act as poisons if they reach too high a concentration. If you wait too long to subculture a microorganism, you risk having a population crash wherein the vast majority of the cells either die or enter a resting state from which it may or may not be trivial to wake them. "Too long" could be as little as 48 hours for some species and a matter of 6+ months for others. The microbiologist must be attuned to the appearance of their microbial cultures and act accordingly.

A key consideration during culture maintenance is that you maintain aseptic technique throughout. Every time you open your microbial culture, you risk contaminating it with one or more other organisms; therefore, you should keep your subculturing routine to the minimum required for your purposes. You must also always be practicing aseptic technique and resist the temptation to get complacent and let your brain switch to autopilot.

16

Characterizing Unknown Organisms

The characterization and identification of unknown microorganisms have been major foci within microbiology, medicine, sanitation, and food manufacture/preparation since visual evidence for the existence of microbes was first published in the mid-1600s AD. The delineation began in earnest during the early-mid 1800s AD as the old **theory of abiogenesis** was gradually supplanted by the **theory of biogenesis,** and many medical professionals began to focus on identifying the material causes of infectious disease.

Traditionally, much of the characterization and identification of microorganisms has been based on microscopic appearance, the manner in which they are stained by various dyes, or how they responded to or changed the appearance of various types of nutrient media. These techniques have the benefits of more than a century's use, rapidly generating data, and in most cases, being less expensive than more modern genomic techniques. These more "low tech" methods still enjoy considerable use in clinical applications where speed of data acquisition, high sample throughput, and per sample cost are essential.

The classification/delineation of organisms is called **taxonomy** which comes from the Greek *taxis* "arrangement" and *–onomy* "method." Humans have been doing taxonomy of plants and animals for as long as we have a history (and probably before). The current standard for bacterial taxonomy is *Bergey's Manual of Systematic Bacteriology, 2nd Edition,* which is a 5-volume (8-part) set of books that delineates bacteria into 33 groups rather than the traditional arrangement of phylum, class, order, and family. In the two most recent editions, the focus of the work has begun to shift away from "artificial" classification systems based on **phenotype** and toward what the authors call "natural" systems based on **genotype (phylogeny)**. Even so, it is acknowledged in the opening chapters of Vol. 1 that classifying bacteria genotypically is fraught with difficulty, especially as we try to classify the higher orders of the taxonomic hierarchy. It should therefore not be surprising when you learn that phenotypic taxonomy still plays a essential role in the taxonomy laid out in *Bergey's* and in microbiology in general.

In this laboratory course, we will be following in the footsteps of the brave men and women who have blazed the trail for us, and we will do so by employing phenotypic classification methods. These are the methods most commonly used in the practice of medicine, and they are the ones that you are most likely ever to come into direct contact with in whatever career you find yourself in.

We will use several aspects of microbial phenotype including colony morphology, cellular morphology, cell wall structure, physiology, and biochemistry to classify a set of bacterial organisms. By this point, you have developed sufficient knowledge of aseptic technique, sample preparation, staining, and specialized growth media that you are ready to move to the next challenge: *Identifying a bacterium whose identity you don't already know.*

A common way of approaching this type of problem is to evaluate the organism's compliment of **enzymes.** Enzymes are arguably the one class of molecule that allows life to exist. They are mostly a subset of the proteins, and they make it possible for many chemical reactions to occur and at temperatures at which life can exist. Different organisms may have different enzymes and may therefore be capable of different types of **cellular metabolism.** There are two major types of enzymes:

- **Intracellular/endoenzymes** – These function inside cells and are responsible for the synthesis of new cell components and the extraction of energy.

- **Extracellular/exoenzymes** – These are made inside the cell but function *outside* the cell. We will look specifically at **hydrolytic exoenzymes** whose function is to use water to break useful high molecular weight compounds into smaller bits that can be transported into the cell.

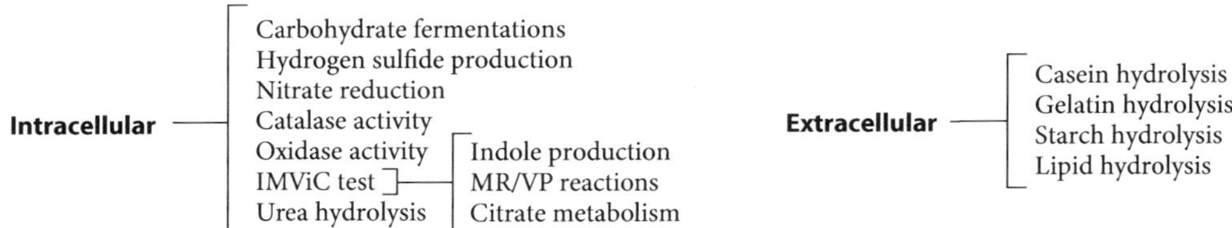

Figure 16-1 | Intracellular vs Extracellular

You will find a list in Table 1 that contains every organism that your instructor will hand out. Beware, however, that not every organism in the table will be handed out this semester. Your job will be to, over the next few weeks, subject your given organism to a battery of tests and use the data that you generate to write a paper in which you identify your organism, recount how you came to the given ID, why you are sure it is *this* organism and not another one, and you'll present relevant details about the organism to the reader to provide a context for why your organism matters.

As you work, keep in mind that no single test can, by itself, give you a positive identification. You will need all of the tests, and you should therefore view your dataset as a profile of the organism. Your job will be to compare your organism's profile to the profiles in Table 1 and make a determination about what organism you have.

Table 16-1 | Biochemical Profiles

Organism	Morphology	Gram stain	Lactose	Dextrose	Sucrose	H$_2$S Production	Indole Production	NO$_3$ Reduction	MR Reaction	VP Reaction	Citrate Use	Catalase Activity	Oxidase Activity	Urease hydrolysis	Casein hydrolysis	Gelatin Hydrolysis	Starch Hydrolysis	Lipid Hydrolysis	Hemolysis
			Fermentations																
Enterococcus faecalis	Coccus	+	A	A	A	-	-	-	+	+	-	-	-	-	+	-	-	-	γ
Lactococcus lactis	Coccus	+	A	A	A	-	-	-	+	-	-	-	-	-	+	-	-	-	γ
Micrococcus luteus	Coccus	+	-	-	-	-	-	±	-	-	-	+	-	+	-	S+	-	-	α
Staphylococcus aureus	Coccus	+	A	A	A	-	-	+	+	±	-	+	-	-	-	R+	-	+	α
Alcaligenes faecalis	Coccobacillus	-	-	-	-	-	-	-	-	±	+	+	+	-	-	-	-	-	α
Bacillus cereus	Bacillus	+	-	A	A	-	-	+	-	±	-	+	-	-	+	R+	+	±	β
Bacillus thuringiensis	Bacillus	+	-	A	-	-	-	+	+	±	-	+	+	+	+	R+	+	-	β
Corynebacterium xerosis	Bacillus	+	-	A±	A±	-	-	+	-	-	-	+	-	-	-	-	-	-	β
Escherichia coli	Bacillus	-	AG	AG	A±	-	+	+	+	-	-	+	-	-	-	-	-	-	α
Klebsiella aerogenes	Bacillus	-	AG	AG	AG±	-	-	+	-	+	+	+	-	-	-	-	-	-	α
Klebsiella pneumoniae	Bacillus	-	AG	AG	AG	-	-	+	±	±	+	+	-	+	-	-	-	-	α
Proteus vulgaris	Bacillus	-	-	AG	AG±	+	+	+	+	-	±	+	-	+	-	R+	-	-	α
Pseudomonas aeruginosa	Bacillus	-	-	-	-	-	-	+	-	-	+	+	+	-	+	R+	-	+	β
Pseudomonas fluorescens	Bacillus	-	-	-	-	-	-	+	-	-	+	+	-	-	-	-	-	-	γ
Salmonella typhimurium	Bacillus	-	-	AG±	A±	+	-	+	+	-	+	+	-	-	-	-	-	-	α
Shigella dysenteriae	Bacillus	-	-	A	A±	-	±	+	+	-	-	+	-	-	-	-	-	-	β

Note: ± = variable reaction, AG = acid and gas, S = slow, R = rapid

17 The IMViC Test Series

The **I**dole, **M**ethyl red, **V**oges–Proskauer, **in** **C**itrate series is a group of tests that use specialized growth media for the purpose of further classifying coliform bacteria. While this testing series is particularly good at delineating members of the **Enterobacteriaceae,** it can also be useful outside that context as you will see by the end of this laboratory exercise.

The Enterobacteriaceae are a family of bacteria that share the traits of being **mesophilic,** short, gram–bacilli that do not form endospores. The IMViC testing series was initially designed to assist clinical professionals in classifying samples containing Enterobacteriaceae so they could determine which they were:

- **Normal intestinal flora** – This category would include organisms in the genera *Escherichia* and *Enterobacter,* for instance, which are **saprophytes** within the latter portion of your intestinal tract. While there may be pathogenic strains within some of the species, most are either **commensal** or **mutualistic** with their human hosts.

- **Occasional pathogens** – This could include species within the genera of *Klebsiella* or *Proteus,* some of which can cause mild to serious illness depending on species, strain, whether the host is immune-compromised, and location in the body.

- **Pathogens** – This would include species within the genera *Salmonella, Serratia, Shigella,* and the world-famous *Yersinia.* These genera have more than a few organisms that can be fairly virulent, successfully attacking even hosts that are otherwise in good health.

Like many of the assays that you have completed thus far, the constituent tests of the IMViC series operate on the principle of identifying the activity of certain biochemical pathways or whether certain enzymatic reactions are taking place.

- **Indole** – This is an assay that will help you know something about the type of metabolism that microorganisms engage in. Specifically, not all microbes can metabolize the amino acid **tryptophan.** Some organisms express the **trytophanase** enzyme which is an initial step of the metabolism of this amino acid. One of the products of this hydrolysis is indole, which can be detected by adding Kovac's reagent to SIM agar which has been inoculated and incubated. Kovac's reagent is composed of butanol, hydrochloric acid, and *p*-dimethylaminobenzaldehyde which will complex with indole yielding a cherry/rose color.

- **Methyl red** – Glucose (dextrose) is a substrate that is used by all Enterobacteriacaeae and a large majority of all other microbes as well. Not all organisms use the same dextrose metabolism pathways and so the presence or absence of these pathways is another avenue that we can use to differentiate them. The MR test will allow you to determine whether an organism will ferment dextrose while producing *stable* acid end products. Some organisms like *E. coli* or *S. aureus* will create acid end products that remain acidic as they do dextrose fermentation resulting in a pH of ~4. You will simply add a few drops of the pH indicator methyl red after incubation. Presence of acid will cause the indicator to turn the broth red. If the broth does not change color or becomes slightly yellow after addition, you have little or no acid present.

- **Voges-Proskauer** – The same broth is used in this assay as was used in the MR test. Some microorganisms such as *K. aerogenes* initially produce acidic end products but later enzymatically convert them to non-acid compounds which results in a slightly higher pH (~6) at the end of the assay. This test will allow you to determine whether any of those non-acid end products of dextrose fermentation are present. This assay requires Barritt's reagents A & B which are alcoholic α-naphthol and 40% KOH, respectively. It takes time for the chemistry to take place, so after incubation, add your reagents then observe the result at 15 or 20 minutes to see whether a rosey red color has developed on the surface of the assay. If you have no color change within 20 min, it means your organism doesn't produce non-acid end products during dextrose fermentation.

- **Citrate** – As you have seen, microorganisms exhibit an array of fermentation pathways and substrates. Some have the ability to ferment citrate if no other fermentable substrates (like dextrose or lactose) are present. To do this, microorganisms must first transport the citrate into their cells via **citrate permease.** Once in the cells, citrate is cloven into oxaloacetic acid and acetate via **citrase.** These products can then be converted to pyruvate and CO_2. If we grow our microbes on Simmons' citrate agar, these metabolic reactions will cause the medium to become alkaline as the CO_2 reacts with Na in the medium to become sodium carbonate (Na_2CO_3). This compound is alkaline and will cause the bromthymol blue pH indicator to change from green to a deep blue.

Procedure

1 | Obtain your maintenance culture:

    a | A single MR/VP tube and an empty, sterilized tube.

    b | A Simmons citrate agar slant.

2 | Inoculate the broth in the standard fashion.

3 | Inoculate the slant by spreading a single loop of inoculum on its surface.

4 | Incubate all media at 37° C for 24 hrs.

5 | Turn to the 24 Hour Return section of this manual and follow the directions there.

Indole Test
Medium: SIM agar
Substrate: Trytophan

Reagents
Kovac's

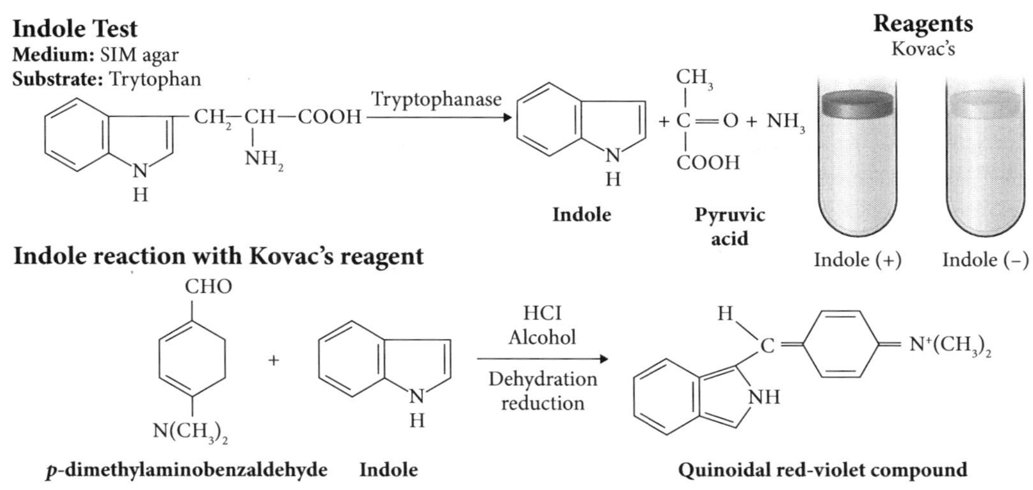

Indole reaction with Kovac's reagent

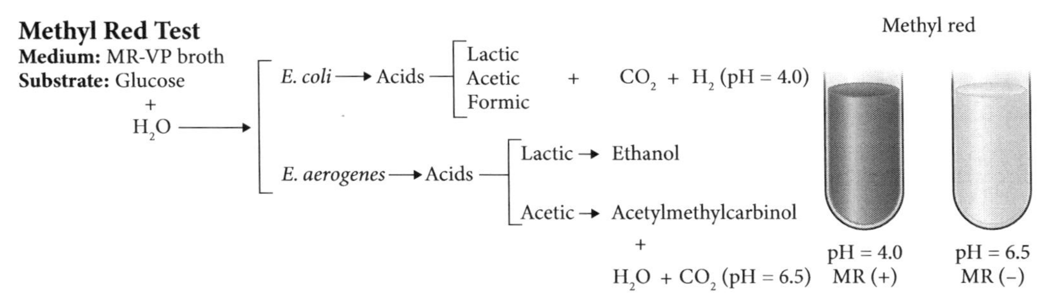

Methyl Red Test
Medium: MR-VP broth
Substrate: Glucose
+
H_2O

Methyl red

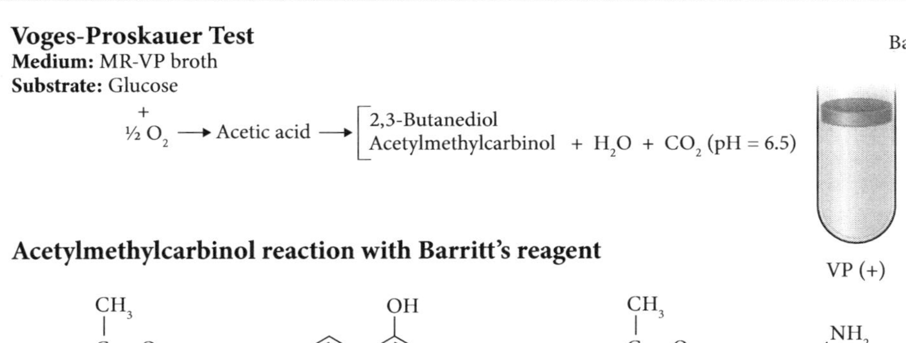

Voges-Proskauer Test
Medium: MR-VP broth
Substrate: Glucose
+
½ O₂ → Acetic acid → 2,3-Butanediol / Acetylmethylcarbinol + H₂O + CO₂ (pH = 6.5)

Barritt's

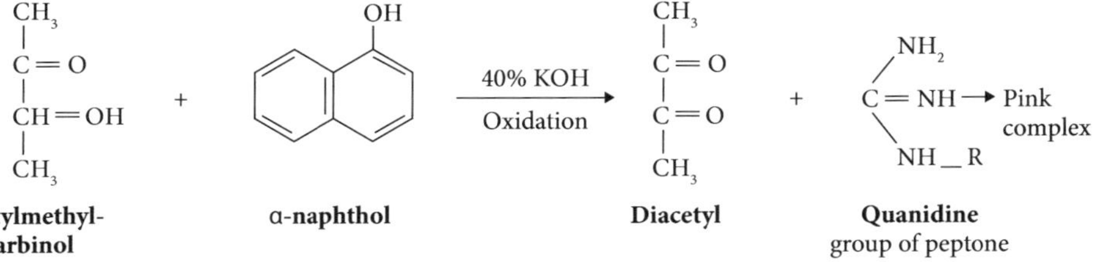

Acetylmethylcarbinol reaction with Barritt's reagent

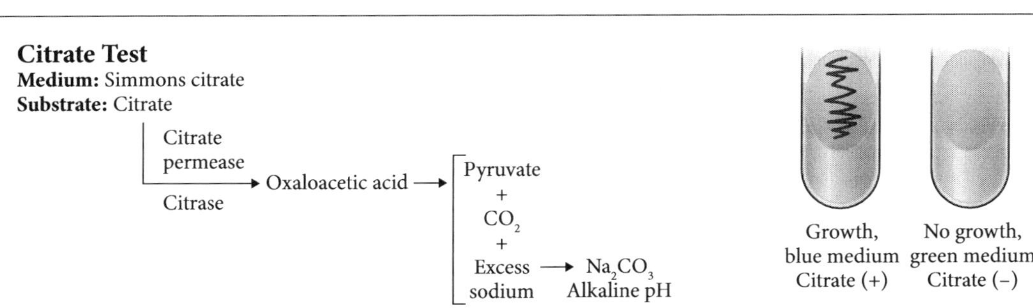

Citrate Test
Medium: Simmons citrate
Substrate: Citrate

Citrate permease / Citrase → Oxaloacetic acid → Pyruvate + CO₂ + Excess sodium → Na₂CO₃ Alkaline pH

Growth, blue medium
Citrate (+)

No growth, green medium
Citrate (−)

Figure 17-1 | IMViC Test Series

Laboratory 18
The EnteroPluri-Test

Clinical laboratories are the place that doctors, physician's assistants, nurses, and nurse practitioners send clinical samples for analysis. These samples may be lavage from the lungs of a pneumonia patient, purulent aspirant from an abscess, urine from a suspected UTI, or a number of other things. Given the number of possible causative organisms, the development of rapid, multi-test systems has been a significant advancement. There are a large number of multi-test systems commercially available and they range from ~\$1/sample to ~\$100/sample depending on the system, level of detailed information, and specificity desired. In this lab, we will be using a test that is ~\$13/sample.

Traditionally, the rapid identification of small, gram- bacilli was one of the more time-consuming tasks in a clinical lab. That task got easier approximately 50 years ago when the Enterotube test was invented (Titsworth *et al.* 1969). This test, recently renamed the EnteroPluri-Test, is a self-contained testing system consisting of an inoculating needle that passes through 12 compartments, each of which contains a different microbiological medium. The 12 different media, along with the post-incubation addition of two external reagents, will allow the user to evaluate a bacterium for 15 different biochemical characteristics. The system is designed to identify members of the *Enterobacteriaceae* and some other gram–, oxidase-negative bacteria.

While the test in intended to ID a certain set of bacteria, it may be used outside of design parameters and still yield some useful results. Pay attention to what is tested in each compartment and what your results are because you may find them useful later when identifying your unknown organism.

Finally, be advised: The EnteroPluri-Test is accurate only if you observe it within the given observation time window. Be sure to refer to the manufacturer's instructions to determine what that time-frame is.

The EnteroPluri Test can be used to identify aerobic, facultatively anaerobic, gram-negative rods (*Enterobacteriaceae*) isolated from any clinical specimen. Examples include: urine, blood, stool, sputum, spinal fluid, and exudate from wounds.

Before using the EnteroPluri Test, the specimen should be cultured to obtain well isolated colonies on general purpose plating media such as MacConkey (MAC), Eosin Methylene Blue (EMB), or Tryptic Soy (TSA) agars. If all colonies plate share the same morphology, only one EnteroPluri Test is needed; if two or more types are present, an additional tube will be needed for each type. The EnteroPluri Test should only be used on oxidase-negative colonies since oxidase-positive organisms (non-*Enterobacteriaceae*) require the use of other media for their further identification.

Limitations of the Product

1 | The EnteroPluri Test is specifically designed to identify gram-negative bacilli which are oxidase-negative and should only be used to differentiate members of the *Enterobacteriaceae* family. Additional biochemical tests are required for identification of other gram-negative bacteria.

2 | The EnteroPluri Test code numbers cannot be used to establish phenotypic identity between isolates from the same or different specimens.

3 | Biochemical results obtained with the EnteroPluri Test may differ from other methods and published material.

4 | *Salmonella* and *Shigella spp.* require serological testing for confirmation. Isolates of these organisms should be referred to state health departments for complete serotyping and epidemiological surveillance.

5 | Identification of *Enterobacteriaceae* should be made with the consideration of additional characteristics such as source of specimen, history of the patient, colonial and microscopic morphology, serology, and antimicrobial susceptibility patterns.

6 | Identification of infrequently isolated/rare isolates should be repeated or additional testing performed to verify the identification of such organisms.

7 | Some strains of organisms may have atypical biochemical reactions due to unusual nutritional requirements or mutations and may be difficult to identify. Some organisms may require longer than 18 h incubation for proper identification

Table 18-1 | Limitations of the Product

	Negative	Positive	Compartment 1
Glucose	Red	Yellow	**Glucose (GLU):** The end products of bacterial fermentation of glucose are either acid, or acid and gas. The shift in pH due to the production of acid is indicated by change in the color of the indicator in the medium from red (alkaline) to yellow (acidic). Any degree of yellow should be interpreted as a positive reaction; orange should be considered negative.
Gas Production	Wax Not Lifted	Wax Lifted	**Gas production (GAS):** This is evidenced by definite and complete separation of the wax overlay from the surface of the glucose medium, but not by bubbles in the medium. Since the amount of gas produced by different bacteria varies, the amount of separation between medium and overlay will also vary with the strain being tested.

	Negative	Positive	Compartment 2
Lysine	Yellow	Purple	**Lysine decarboxylase (LYS):** Bacterial decarboxylation of lysine, which results in the formation of the alkaline end product cadaverine, is indicated by a change in the color of the indicator in the medium from pale yellow (acidic) to purple (alkaline). Any degree of purple should be interpreted as a positive reaction. The medium remains yellow if decarboxylation of lysine does not occur.

	Negative	Positive	Compartment 3
Ornithine	Yellow	Purple	**Ornithine decarboxylase (ORN):** Bacterial decarboxylation of ornithine, which results in the formation of the alkaline end product putrescine, is indicated by a change in the color of the indicator in the medium from pale yellow (acidic) to purple (alkaline). Any degree of purple should be interpreted as a positive reaction. The medium remains yellow if decarboxylation of ornithine does not occur.

	Negative	Positive	Compartment 4
H$_2$S	Beige	Black	**H$_2$S production (H$_2$S):** Hydrogen sulfide is produced by bacteria capable of reducing sulfur-containing compounds, such as peptones and sodium thiosulfate present in the medium. The hydrogen sulfide reacts with the iron salts also present in the medium to form a black precipitate of ferric sulfide usually along the line of inoculation. Some Proteus and Providencia strains may produce a diffuse brown coloration in this medium, however, this should not be confused with true H$_2$S production, i.e., presence of black color. **NOTE:** The black precipitate may fade or revert back to negative if EnteroPluri Test is read after 24 h incubation.

	Negative	Positive	Compartment 5
Adonitol	Red	Yellow	**Adonitol (ADON):** Bacterial fermentation of adonitol, which results in the formation of acidic end products, is indicated by a change in color of the indicator present in the medium from red (alkaline) to yellow (acidic). Any sign of yellow should be interpreted as a positive reaction; orange should be considered negative.

	Negative	Positive	Compartment 6
Lactose	Red	Yellow	**Lactose (LAC):** Bacterial fermentation of lactose, which results in the formation of acidic end products is indicated by a change in color of the indicator present in the medium from red (alkaline) to yellow (acidic). Any sign of yellow should be interpreted as a positive reaction; orange should be considered negative. This test is useful to confirm the lactose reaction of colonies taken from various enteric differential media, e.g., Salmonella Shigella (SS), MacConkey (MAC), Eosin Methylene Blue (EMB).

	Negative	Positive	Compartment 7
Arabinose	Red	Yellow	**Arabinose (ARAB):** Bacterial fermentation of arabinose, which results in the formation of acidic end products, is indicated by a change in color of the indicator present in the medium from red (alkaline) to yellow (acidic). Any sign of yellow should be interpreted as a positive reaction; orange should be considered negative.

	Negative	Positive	Compartment 8
Sorbitol	Red	Yellow	**Sorbitol (SORB):** Bacterial fermentation of sorbitol, which results in the formation of acidic end products, is indicated by a change in color of indicator present in the medium from red (alkaline) to yellow (acidic). Any sign of yellow should be interpreted as a positive reaction; orange should be considered negative.

	Negative	Positive	Compartment 9
Voges-Proskauer	Colorless	Red	**Voges-Proskauer (VP):** Acetylmetylcarbinol (acetoin) is an intermediate in the production of butylene glycol from glucose fermentation. The production of acetoin is detected by the addition of two drops of a 20% w/v aqueous solution of potassium hydroxide containing 0.3% w/v of creatine and three drops of a 5% w/v solution of alphanaphthol in absolute ethyl alcohol. The presence of acetoin is indicated by the development of a red color within 20 min. However, most positive reactions are evident within 10 min.

	Negative	Positive	Compartment 10
Dulcitol	Green	Yellow	**Dulcitol (DUL):** Bacterial fermentation of dulcitol, which results in the formation of acidic end products, is indicated by a change in color of the indicator present in the medium from green (alkaline) to yellow or pale yellow (acidic).
PA	Green	Black-smokey-gray	**Phenylalanine deaminase (PA):** This test detects the formation of pyruvic acid from the deamination of phenylalanine. The pyruvic acid formed reacts with a ferric salt present in the medium to produce a characteristic black gray to smoky gray color. Ferric chloride need not be added since the medium already contains an iron salt. The medium may become a deeper shade of green after incubation. This should be considered negative.

	Negative	Positive	Compartment 11
Urea	Beige	Red-purple	**Urea (UREA):** Urease, an enzyme possessed by various microorganisms, hydrolyzes urea to ammonia causing the color of the indicator in the medium to shift from yellow (acidic) to red-purple (alkaline). The urease test is strongly positive for Proteus species and may be evident as early as 4 to 6 h after incubation; it is weakly positive (light pink color) after 18 to 24 h incubation for Klebsiella and Enterobacter species.

	Negative	Positive	Compartment 12
Citrate	Green	Blue	**Citrate (CID):** This test detects those organisms which are capable of utilizing citrate, in the form of its sodium salt, as the sole source of carbon. Organisms capable of utilizing citrate produce alkaline metabolites which change the color of the indicator from green (acidic) to deep blue (alkaline). Any degree of blue should be considered positive. **NOTE:** Certain microorganisms will not always produce the ideal "strong" positive color change. Lighter shades of the same basic color should also be considered positive.

NOTES

- A red or orange color in a carbohydrate compartment is a negative reaction; only yellow is considered a positive reaction.

- *Proteus vulgaris* may give a weak positive reaction in the glucose compartment. This is acceptable.

- The amount of gas produced by *Proteus vulgaris* and *Escherichia coli* may be weak and at times a negative reaction may be observed. This is acceptable.

- The dulcitol reaction for *Escherichia coli* and *Klebsiella pneumoniae* may be weak or negative after 24 h incubation. This is acceptable.

- The sorbitol reaction for *Escherichia coli* may be weak or at times negative. This is acceptable.

- The urea reaction for *Pseudomonas aeruginosa* may be weak or negative after 24 h incubation. This is acceptable.

- Arabinose reaction for *Pseudomonas aeruginosa* may be positive after 18 to 24 h incubation. This is acceptable.

Procedure

1 | Obtain your maintenance culture:

    a | An EnteroPluri-Test.

2 | Inoculate the EntroPluri-Test according to the TA's instructions.

3 | Incubate at 37° C for 24 hrs.

4 | Turn to the 24 Hour Return section of this manual and follow the directions there.

    a | Be sure to observe the Enteropluri-Test within the specified time frame. Premature or delayed observations will invalidate results.

Table 18-2 | EnteroPluri-Test with VP

DATE						SAMPLE								
	Group 1			Group 2			Group 3			Group 4			Group 5		
Test	Glucose	Gas	Lysine	Omithine	H₂S	Indole	Adonitol	Lactose	Arabinose	Sorbitol	VP	Dulcitol	PA	Urea	Citrate
Positivity Code	4	2	1	4	2	1	4	2	1	4	2	1	4	2	1
Results															
Code Sum															
NUMERICAL CODE						MICROORGANISM									

References

Becton Dickinson & Co. (1999). Enterotube II, Identification system for enterobacteriacaea. Becton Dickinson Microbiology Systems, catalog no. 211832.

Titsworth E, Grunberg E, Beskid G, Cleeland R, & Delorenzo, WF (1969). Efficiency of a multitest system (Enterotube) for rapid identification of Enterobacteriaceae. Applied microbiology, 18(2), 207–13.

Nitrate Reduction: A Type of Anaerobic Respiration

Fermentations aren't the only type of metabolism that we can use to classify organisms. There is more than one type of respiration, for instance. Organisms might be aerobic or they might be anaerobic or they may be aerotolerant. The types of respiration you find operating in organisms within those categories are different and may therefore be useful for classification. In this laboratory exercise, you will be investigating the ability of some organisms to use nitrate (NO_3^-) as a source of terminal-electron-accepting oxygen.

Nitrate reduction has more than one step, and not all organisms do all the steps. Hence, there are partial and complete NO_3^- reductions. For the purposes of this lab, we will define them as follows:

Partial – Ionic product that stays in solution

$$NO_3^- + 2\,H^+ + 2\,e^- \rightarrow NO_2^- + H_2O \text{ (Nitrate reductase)}$$

Complete – Gaseous product that escapes the solution

$$NO_2^- + 2\,H^+ + e^- \rightarrow NO + H_2O \text{ (Nitrite reductase)}$$
$$2\,NO + 2\,H^+ + 2\,e^- \rightarrow N_2O + H_2O \text{ (Nitric oxide reductase)}$$
$$N_2O + 2\,H^+ + 2\,e^- \rightarrow N_2 + H_2O \text{ (Nitrous oxide reductase)}$$

It isn't uncommon for microorganisms to have both aerobic and anaerobic respiration pathways and to use whichever one is more apt for the given circumstances. All the organisms that we use in this lab are capable of growth in the presence of atmospheric oxygen (O_2), but they may be **obligate aerobes, facultative aerobes,** or **aerotolerant** (Refer to Laboratory 28).

This matters because among the organisms we use in this lab, it is only the facultative anaerobes who may do NO_3^- reduction. Facultative anaerobes will preferentially use O_2 via their aerobic pathway if O_2 is available. By growing our microbes in **trypticase nitrate broth** that has just enough agar added to it to make it semi-solid, we slow the diffusion of O_2 into the medium. This means that any organisms with an aerobic pathway will use up the O_2 at a faster rate than it can diffuse back in. This will create an anaerobic environment extending from just below the medium surface to the bottom of the tube. In response, any facultative anaerobes will change from their aerobic respiration to anaerobic respiration. Once that change-over occurs, any organisms that can reduce NO_3^- will begin to do so.

We will use two (possibly three) reagents to help us determine whether our organisms are capable of doing nitrate reduction. **Reagent A** is **sulfanilic acid** and **Reagent B** is **α-napthylamine,** and depending on your initial result, you may need to use **zinc powder.** Reagents A and B are colorless liquids that together will react with any nitrous acid (HNO_2) present in the medium to form a deep red color. Zinc is capable of catalyzing the reduction of NO_3^- to NO_2^- all by itself.

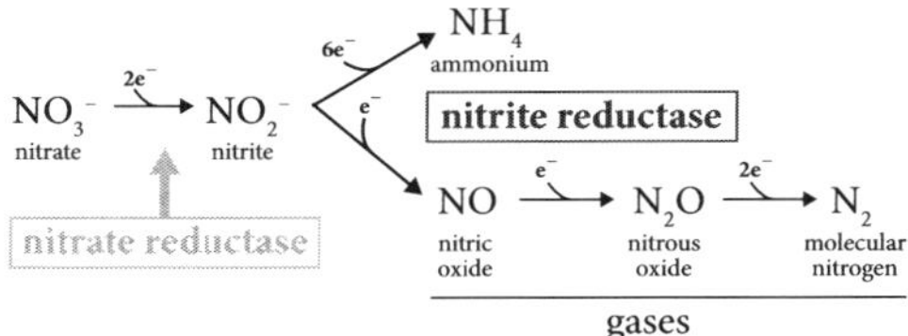

Figure 19-1 | Nitrate Reduction and Denitrification

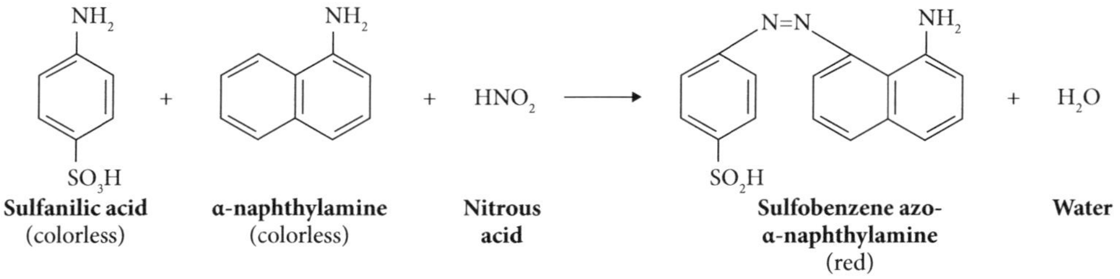

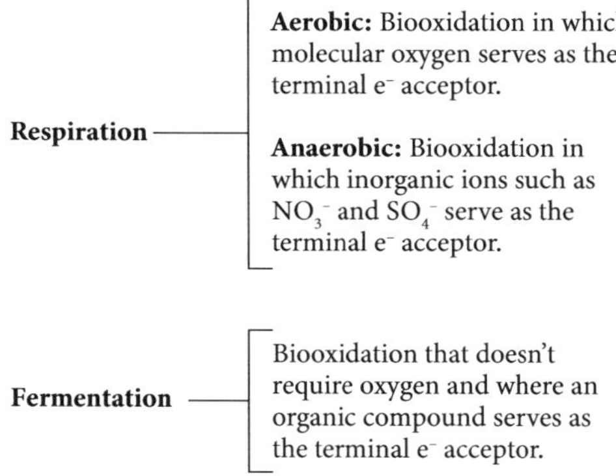

Figure 19-2 | Formation of Colored Complex Indicative of NO_3^- Reduction

Procedure

1 | Obtain a tube of trypticase nitrate broth for every organism you are to investigate.

2 | Inoculate your media with those organisms by transferring one loop of the organism into the broth.

3 | Incubate at 37° C for 24 hrs.

4 | After incubation, add 5 drops of Reagent A followed by 5 drops of Reagent B.

5 | If the reagents bring about a red color within 30 seconds, your experiment is done.

6 | If the reagents bring about no color change, add a small scoop of zinc powder and watch for any development of a red color.

Figure 19-3 | Respiration vs Fermentation

Laboratory 20
Sugar Fermentations

As we've discussed before, fermentations can be diagnostic for identifying microorganisms. In this exercise, you will determine whether some microorganisms can or cannot ferment two carbohydrates: dextrose and sucrose.

The ability of organisms to utilize **catabolic compounds** is determined by that organism's complement of enzymes. A very common catabolic pathway is **glycolysis** whose end products still contain a significant amount of energy. Some microorganisms are capable of fermenting these glycolytic end products to keep their co-enzymes (like NAD+) free to continue facilitating more glycolysis. Microorganisms may do this using an aerobic pathway, others use an anaerobic pathway, and still others may use either depending on the situation, and some organisms do not do this at all. Each pathway has specific products, some of which, such as acids and gasses, are readily diagnostic.

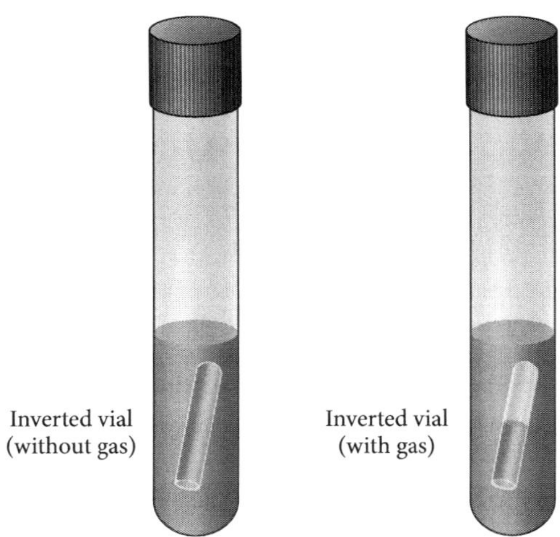

Figure 20-1 | Durham Tube Inside a Standard Test Tube

In this exercise, you will be using standard test tubes containing a sugar-fortified nutrient broth, a pH indicator, and a Durham tube. The nutrient broths will contain either dextrose or sucrose along with phenol red which has a red/purple color at a neutral pH and turns yellow in the presence of acid. The **Durham tube** is simply a small test tube that has been placed inverted in the broths. If any of the organisms ferment glycolytic products and thereby produce one or more gasses, the Durham tube will capture a fraction of the gasses forming a gas bubble.

Be advised that an absence of detectable fermentation doesn't mean the organism didn't grow in the medium. There are other energy-containing compounds in the nutrient mix such as peptones, the processing of which produces ammonium hydroxide (NH_4OH) which is alkaline.

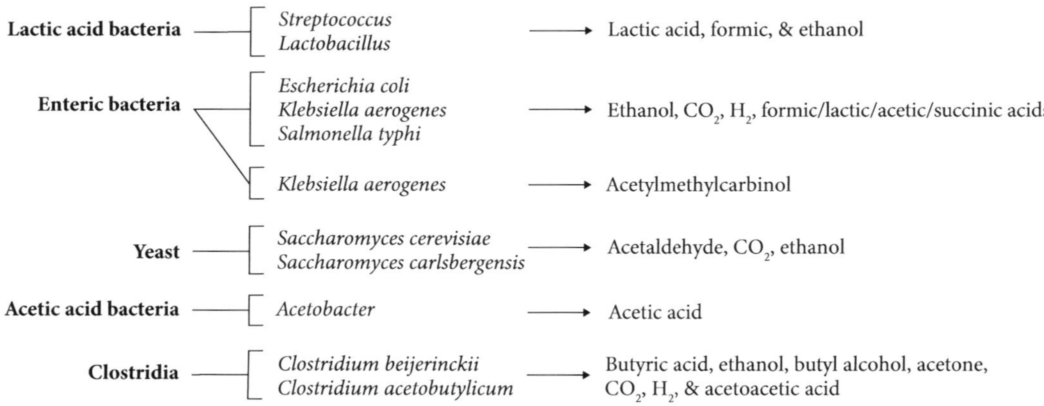

Figure 20-2 | *Examples of Sugar Fermentation End-Products.*

Procedure

1 | Obtain one of each broth for every organism you are investigating.

2 | Inoculate the broths with a single loop of microbial culture. Be careful not to stir up the broth as you inoculate as you may introduce air bubbles into the Durham tube.

3 | Incubate at 37° C for 24 hrs.

4 | After incubation, observe the color of the broths and whether the Durham tube captured any gasses. You should see no gas production in tubes that have not acidified.

Cytochromes are iron-containing hemeproteins that are critical to **electron transport chains (ETCs).** Various microorganisms may use various ETCs within which are various cytochrome compounds which function as e^- carriers. To be effective e^- carriers, cytochromes must be repeatedly reduced and then oxidized. Interruption of these reduction/oxidation loops is a type of poisoning which can lead to organism death. The oxidation part of the loop is accomplished via **cytochrome oxidase enzymes.**

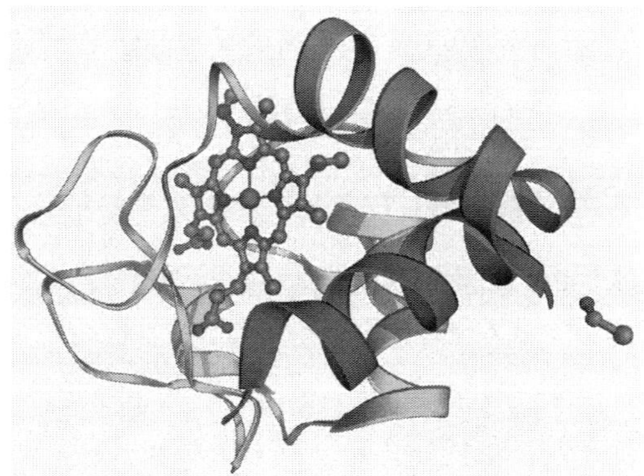

Figure 21-1 | Cytochrome C

© Laguna Design/Science Source

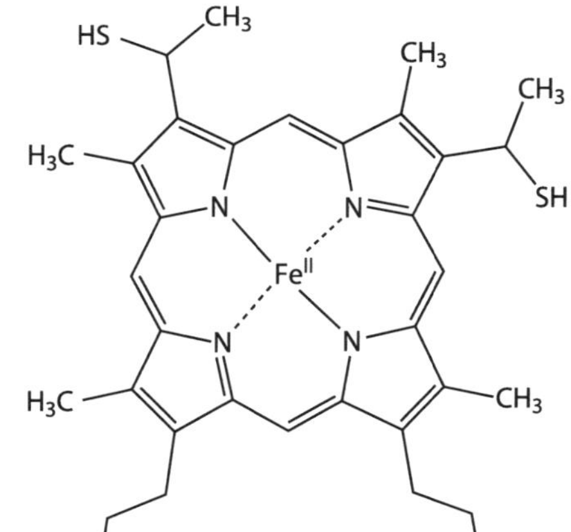

Figure 21-2 | Heme C

As you may deduce from its name, cytochrome oxidase catalyzes the oxidation of a reduced cytochrome using molecular oxygen (O_2). This oxidation produces H_2O and hydrogen peroxide (H_2O_2). Aerobic bacteria, some facultative anaerobes, and some microaerophiles exhibit cytochrome oxidase activity. The ability isn't universal and can therefore be used for microbial classification. For instance, it is a fairly quick method of differentiating between the small, gram– bacilli of the Enterobacteriaceae (oxidase negative) and the small, gram– bacilli of the genera *Neisseria* and *Pseudomonas* (oxidase positive).

The expression of cytochrome oxidase can be detected via the addition of tetramethyl-p-phenylenediamine which is initially colorless but will turn a deep blue-black in the presence of cytochrome oxidase. This is the method you will use to classify bacteria during this laboratory exercise.

Procedure

Obtain plates of the organisms you are to investigate and a dropper bottle of tetramethyl-p-phenylenediamine.

Direct Method

1 | Place ***one drop*** of the reagent directly onto a single colony in your plate, ***being careful not to touch the nozzle of the bottle to the colony***. If you touch the colony, you will contaminate the bottle and ruin the rest of the reagent.

2 | Watch for a change of the reagent from colorless to blue-black (usually within 10 seconds).

3 | This method should not be used with colonies on blood agar, as the agar is likely to contain active oxidizers within it, which can yield false-positive results. For colonies grown on blood agar, use the filter paper method.

Filter Paper Method

1 | Obtain sterilized filter papers.

2 | Place ***one drop*** of tetramethyl-p-phenylenediamine on a filter paper.

3 | Using your inoculating loop, transfer a colony from your plate to the reagent-wet filter paper.

4 | Watch for a color change to blue–black within 10 seconds.

If you use the filter paper method, beware of using reactive wire loops such as NiChrome. Generally, the use of a loop constructed of anything other than an inert metal such as platinum (Pt) will cause spontaneous oxidation of the reagent. Thankfully, this loop-induced oxidation is typically slower than metabolic oxidation, taking place after ~10 seconds.

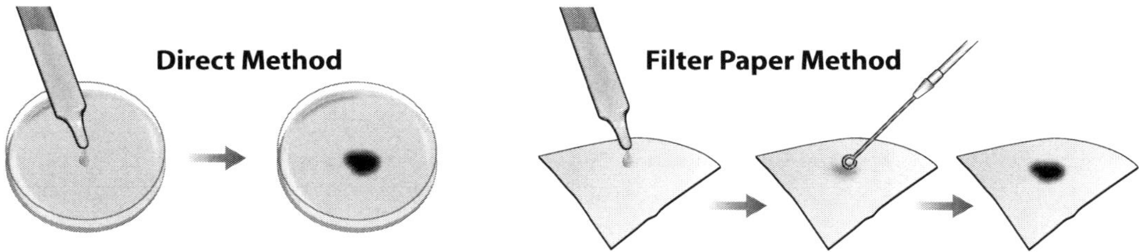

Figure 21-3 | The Two Methods of Executing the Oxidase Test

22

Laboratory 22
Catalase Test

Not all the byproducts of aerobic metabolism are harmless to the cell. Some can be extremely harmful, such as highly-reactive hydrogen peroxide (H_2O_2) or superoxide (O_2^-). If not converted to something less reactive, these strong oxidizers will rampage around inside the cell oxidizing proteins, enzymes, DNA, RNA, and all manner of things. O_2^- is so toxic that it is used by the human immune system as a means of killing microbial invaders. It is therefore detrimental for a microbial cell to generate these compounds as part of their standard metabolism and to have no means of rendering them less harmful. These methods of catalytically converting toxic compounds to less-harmful by-products are not universal, so they can be used to classify microorganisms.

In this laboratory exercise, you will be investigating whether various microorganisms express **catalase** which catalyzes the breakdown of $2H_2O_2 \rightarrow 2H_2O + O_2$. Not all aerobic organisms express catalase, and those that don't do still express **superoxide dismutase** which catalyzes $O_2^- \rightarrow H_2O_2$ which, while still toxic, is much less toxic than superoxide. Anaerobic organisms do not express catalase, peroxidase, or superoxide dismutase and therefore cannot survive in environments with significant amounts of O_2.

Procedure

1 | Obtain the plates of your given microorganisms, a bottle of diluted H_2O_2, and a dropper.

2 | Place *one drop* of the dilute H_2O_2 directly onto a colony on your plate.

3 | Watch for immediate bubbling and frothing of the liquid which indicates a positive test.

Extracellular Enzymatic Activities

Some useful molecules are too big to bring into the cell. There are plenty of high-molecular-weight molecules that could be useful, if only a cell could manage to get them imported. Polysaccharides, lipids, and proteins all fit this description. One way that a microbial cell may tap into these resources is to produce and secrete **extracellular enzymes** or **exoenzymes.**

Exoenzymes are produced by many organisms, usually gram+ bacilli. Most of these enzymes are **hydrolases** that use water in the process of splitting the bonds that tie together large macromolecules.

In this laboratory exercise, you will investigate whether several microorganisms produce exoenzymes capable of hydrolyzing a few macromolecules:

Starch

This is a branching polymeric carbohydrate (polysaccharide) that consists of a large number (poly–) of glucose molecules (–mers) covalently bonded together by **glycosidic bonds.** These bonds can be hydrolyzed by **amylases,** which split the large starch molecules into smaller chunks. The smaller chunks are **dextrins** and **maltose.** Dextrins will form spontaneously if you expose starch to high heat and are partially responsible for the browning of toast. Dextrins are split until you are left with molecules of **maltose,** which consist of two glucose molecules bound together. Finally, maltose is hydrolyzed by **maltase** which yields soluble glucose that can be brought into the cell.

You will be using agar that includes 8% starch by dry weight. This type of agar is a translucent white, and it often continues to look that way even if you grow an organism that expresses amylases. When you expose starch to a tincture of iodine, the iodine will temporarily stain the starch black. Hence, you will use a tincture of iodine to show you whether any of the organisms hydrolyzed the starch. If any of them do, the media immediately adjacent to the microbial growth will remain translucent in the presence of the iodine while the medium further away will turn black.

Gelatin

Gelatin is an incomplete protein because it lacks tryoptophan, but that doesn't stop microorganisms from using it as a nutrient. Proteins are polymers of amino acids where the amino acids are covalently bonded by **peptide bonds.** Gelatin is formed via hydrolysis of collagen. Gelatin is solid below 25° C, and above that, it will liquefy. Some organisms are capable of producing **gelatinase** which will hydrolyze gelatin. That hydrolysis causes the medium to liquefy, and it will remain liquid at room temperature and even down to temperatures near the freezing point of water. The value of gelatin for identifying bacteria has been recognized since the late 1800's.

You will be using TSB into which 12% gelatin by dry weight has been dissolved. You will stab inoculate the medium in the test tubes and incubate it for 48 hours then place it in the refrigerator until it fully cools down. Any tube that remains liquid after refrigeration indicates that the organism did **rapid gelatin hydrolysis.** Any that are solid may be re-incubated for 5 days and then refrigerated. Tubes that are liquefied after the second incubation indicate that the organism is capable of **slow gelatin hydrolysis.**

Casein

This is one of the components that make milk white, and it is the major protein in milk which is why it is otherwise known as "milk protein." Proteins must be hydrolyzed before they can be brought into a cell, and the hydrolysis proceeds in steps: protein -> peptones -> polypeptides -> dipeptides -> amino acids. Collectively, these steps are **proteolysis,** and it is done by **proteases.**

You will be using a nutrient agar which includes 70% by dry weight instant non-fat dry milk. This agar typically has a milk-like color that is more opaque than starch agar. In this preparation, which excludes milk fats, it is the casein that gives the agar its milky color. You will inoculate your casein agar plate and incubate it as normal. After incubation, if you find that the medium immediately under and adjacent to the microbial growth has become transparent, that indicates that the organism on that plate produced a **caseinase** that hydrolyzed the casein in the medium.

Lipids

Lipids, also known as fats, are large macromolecular compounds that are often used to store a tremendous amount of energy. Lipids are a class of compounds encompassing a number of different basic structures. One common subclass of lipids is **triglycerides,** which consist of a glycerol cap attached to three fatty acid chains. These are initially hydrolyzed by a class of **lipases** called **esterases** which split the fatty acid tails off the glycerol cap. Once split off, the fatty acids and the glycerol can be brought into the cell and further broken down so as to be run through the Krebs Cycle.

You will be using a nutrient agar that includes 30% tributyrin by dry weight. The presence of the high concentration of lipid causes the otherwise clear agar to become somewhat cloudy. After inoculating and incubating in the standard way, if you find that the medium immediately under and adjacent to the microbial growth has become transparent, that indicates that the organism in that plate produced a lipase capable of hydrolyzing the **ester** bonds binding the fatty acids to the glycerol.

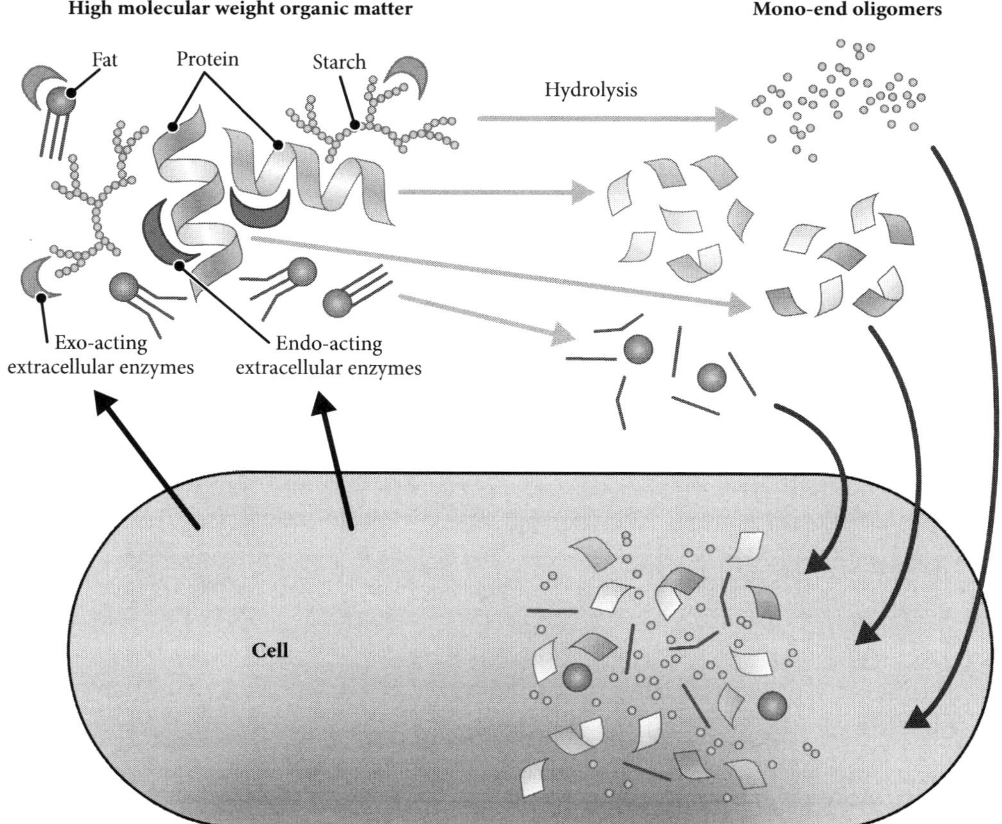

Figure 23-1 | Extracellular Enzymatic Activities

Procedure

1 | Obtain your maintenance culture and a medium of each type.

2 | Inoculate your organism onto each agar plate with a standard 3-way streak method.

3 | Inoculate your organism into the gelatin tube by stabbing once into the center of the gelatin surface.

4 | Incubate the agar plates at 37° C for 24 hrs.

5 | Incubate the gelatin tube at 37° C for 48 hrs (rapid method).

Differential, Selective & Enriched Media

All of the various microbiological growth media that we have used this semester have been **nutrient media.** That is, they have the basic nutrients that at least *some* microbes need for growth and reproduction. Some of the media that we have used are more than just nutrient media, though. These special-purpose media have additional ingredients or perhaps lack specific components. Some of these media are **selective media** and some of them are **differential media** and some do both jobs at once.

Selective media are designed to inhibit the growth of some types of microorganisms while allowing the growth of other types. These "types" may encompass categories like bacteria vs. fungi (TSA vs. Sabouraud agar) or gram+ vs. gram– (phenyethyl alcohol agar vs. MacConkey agar). In this laboratory course, you will use one selective medium:

- **Phenylethyl alcohol agar (PeOH)** – Contains 6% phenylethyl alcohol (a.k.a **phenethyl alcohol**) which is somewhat inhibitory to most gram– organisms and to a few gram+ organisms. This medium is used to isolate gram+ organisms, especially if your culture has become contaminated with swarming *Proteus* species. If an organism's survival and reproduction are inhibited, then its colonies and colony morphology should be smaller and thinner than they would be on a non-selective medium.

Differential media are designed to change in color or clarity in response to certain types of microbial biochemistry. The ability to ferment certain carbohydrates and produce acidic end products is a common analytical category. Change out the carbohydrates or the pH indicator and you have a different type of differential medium. Add in some ingredients that inhibit certain microorganisms and you have a medium that is both differential and selective. In this laboratory course, you will use the following media that are simultaneously differential and selective:

- **Mannitol salt agar (MSA)** – This is a medium that it commonly used in clinical settings to isolate *Staphylococcus* species. It is able to do this because the medium contains 7.5%–10% NaCl which is a relatively high concentration of salt that inhibits the growth of most clinical bacteria outside of genus *Staphylococcus.* Moreover, it can be used to rapidly identify which are likely *Staphylococcus aureus* because the medium contains the carbohydrate mannitol and the pH indicator phenol red which together allow the differentiation of colonies capable of fermenting mannitol (common in *S. aureus*) and those that are not. The medium is initially red, and if the organism produces acid (the byproduct of mannitol fermentation), the medium will turn yellow.

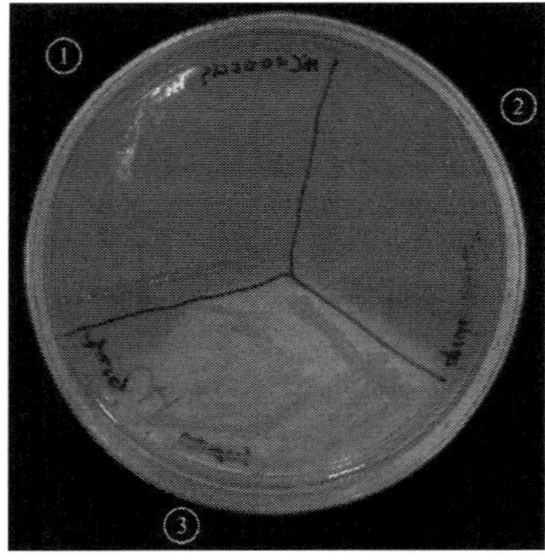

Figure 24-1 | MSA. After 24 hours, this inoculated Mannitol Salt Agar (MSA) culture plate cultivated colonial growth of: 1) *Micrococcus* sp; 2) *S. epidermidis;* 3) *S. aureus*

- **MacConkey agar (MCA)** – This medium allows for the isolation of gram– organisms and for the differentiation between those that can ferment lactose while producing acid and those that cannot. Organisms that are gram– and ferment lactose are presumed to be **coliform bacteria.** Gram– organisms that do not ferment lactose are presumed to be **typhoid, paratyphoid, or dysentery bacteria.** The medium contains the dye crystal violet and bile salts which are inhibitory to gram+ organisms and fungi. Also present are the carbohydrate lactose and pH indicator neutral red which together allow an investigator to determine whether there are organisms present capable of fermenting lactose and producing an acid as a byproduct. The medium is initially a light pink/red color and will remain so if gram– organisms incapable of fermenting lactose are grown on it; the colonies will generally be transparent and the same color as the medium. If an organism is capable of lactose fermentation, then the colonies will turn an opaque, bright pink color. If the organism vigorously ferments the lactose, producing copious amounts of acid end products, the medium immediately adjacent to the colonies will lose clarity and become more opaque as the bile salts precipitate out of the acidified medium and form crystals within the agar matrix.

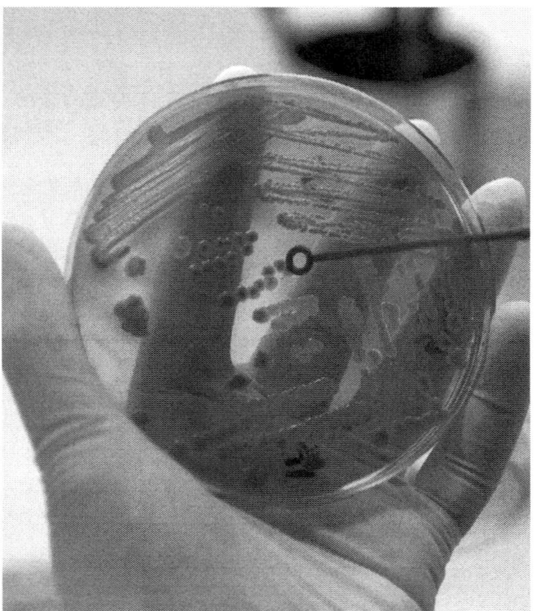

Figure 24-2 | MCA. This is what 18 hrs of growth looks like on an MCA plate inoculated with *E. coli.*

- **Eosin-methylene blue agar (EMB)** – This medium shares a number of similarities to MCA in that it also isolates gram– organisms and allows the investigator to determine if any ferment lactose producing acid. The medium contains the dyes eosin Y and methylene blue which both inhibit gram+ bacteria and will be absorbed by bacterial colonies under acidic conditions. These dyes are not as strongly inhibitory as the combination of CV and bile salts in MCA, so some small growth of gram+ organisms is common on this medium. Also present in the medium is lactose, which may or may not be fermented by the organisms grown on the medium. The medium is initially a transparent purple color, and organisms incapable of lactose fermentation won't change this; they'll be transparent. If an organism is capable of fermenting lactose, then at least the center of the colony will become more opaque as the acidic environment encourages the colony to absorb the dyes. It is common for not just the center but the entire colony to become a dark, opaque blue–black. This medium can be useful in specifically identifying *Escherichia coli* and *Klebsiella aerogenes* whose colonies will be metallic green and a slimy pink color respectively.

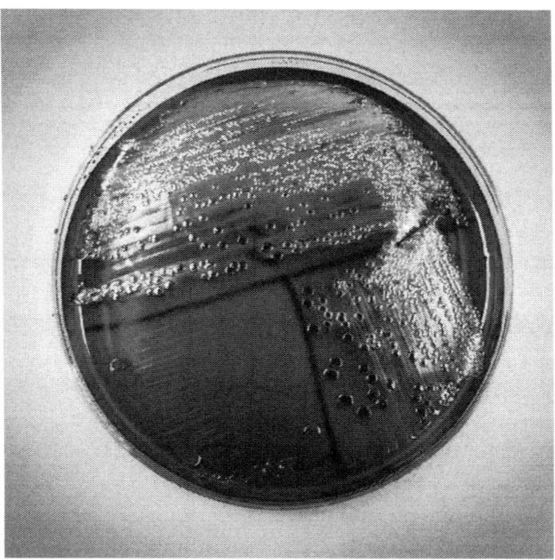

Figure 24-3 | EMB. *E. coli* culture on EMB agar, a selective and differential medium in which the growth of gram-positive bacteria is inhibited. Moreover, those organism who can ferment lactose, such as *E. coli,* will absorb the dye contained on the medium and will have a distinctive metallic green as shown in the picture.

- **Sheep blood agar** – This is technically an **enriched medium** because blood is a very high-nutrient ingredient and contains many unique trace components. Media of this kind permit the growth of **fastidious organisms.** This agar can also function as a differential medium. Specifically, this medium can help an investigator determine whether any of their organisms are capable of performing **hemolysis,** which is the breakdown of red blood cells (RBCs) and their hemoglobin. Hemolytic activity is a **pathogenic indicator** which is caused by the **virulence factor** hemolysin (an enzyme). Therefore, hemolytic activity can be a clinically important factor to identify. Beware, however, that some species, such as some of the Streptococci, express hemolysins that are deactivated by the presence of O_2; therefore it is common practice to stab the blood agar plate in an open spot with your loop after inoculating it. The lower-O_2 concentration within the agar may reveal the presence of these hemolysins. There are three classes of hemolytic activity:

  - **α (alpha)** – Incomplete hemolysis where hemoglobin is partially metabolized to methemoglobin (pronounced "met-hemoglobin") which gives the immediately adjacent medium a green/grayish tinge. Some α hemolyzers will clear the medium under the colonies but retain the green/gray color.

  - **β (beta)** – Complete hemolysis wherein the RBCs are completely lysed and the hemoglobin is more fully metabolized. Organisms capable of this feat will have a zone of transparent medium under and immediately adjacent to colony growth. This transparent zone will also be much paler, no longer red, and will have no green/gray coloration in the medium.

- **γ (gamma)** – No hemolysis. Organisms that exhibit this class of hemolysis are actually not doing hemolysis at all. Colonies will commonly appear a white/gray color and will affect no change in medium color or clarity.

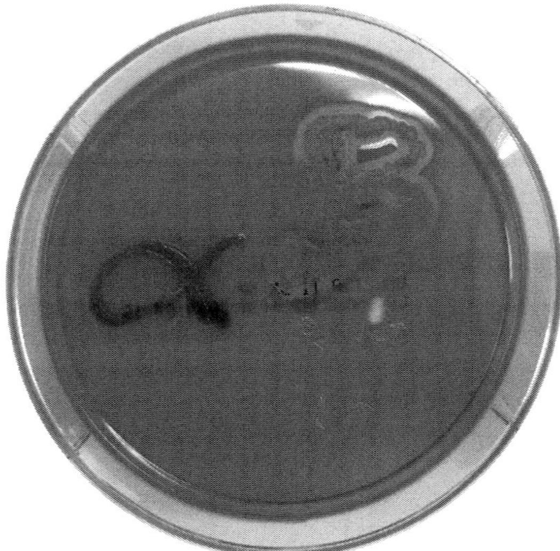

Figure 24-4 | SBA. Petri dish containing sheep's blood agar (SBA) inoculated with *Escherichia coli* (alpha), *Bacillus cereus* (beta), and *Enterococcus faecalis* (gamma). This is the result after 48 hrs of incubation at 37° C. Alpha is a green color, which is bleeding into the medium, beta exhibits a clear zone around the colony growth, and gamma changed nothing about the medium.

Procedure

1 | Obtain your maintenance culture and a plate of each medium.

2 | Inoculate your organism onto the surface of the medium in each plate.

3 | Incubate all plates at 37° C for 24 hrs, except for the blood agar plates which should be incubated for 48 hrs.

4 | After incubation, observe your plate and interpret any changes in the medium or colony appearance according to the principles laid out above.

Laboratory 25
Effects of Temperature on Growth

Microbial growth is defined as an increase in the number of microbial cells over time, and this growth is highly dependent upon environmental temperature. This dependence is founded on the enzyme compliment that a particular microbe has. There are enzymes that work efficiently in temperatures near the freezing point of water, there are enzymes that work well at room temperature, and there are enzymes that can function above the boiling point of water. Thus any talk of "high" or "low" temperature must be relative because whatever temperature is high or low to one organism may be ideal for another and fatal to yet another.

Most **enzymes** are a class of proteins, and like all proteins, they have a 3D structure that is key to their functioning. If that 3D structure is altered, say through an increase in temperature that causes a shuffle in the secondary or tertiary bonding, then the enzyme will no longer function as it previously did. This change in 3D structure is called **denaturation,** and it can be temporary or permanent depending on the severity and duration of the temperature alteration. Permanent denaturation of cellular enzymes results in cell death.

Conversely, cold temperatures do not denature enzymes but rather slow the rate of their chemical reactions. As reaction times are slowed, the interactions between enzymes and substrates are also slowed. The result is a slower-than-optimal reaction rate and a correspondingly slower growth rate. If temperatures fall far enough, enzyme reaction rates can fall to zero, and microbial metabolism stops. This doesn't mean that the microbe is dead, however. If the cell retains its cell wall and membrane integrity, it is perfectly capable of resuming metabolism if the temperature rises.

Bacteria are not the only prokaryotes in the world, but they are the main focus of this laboratory. In bacteria, it is most common to find enzymes that will function somewhere within the range of –5° C to 80° C. It is common for a particular bacterium to have a growth temperature range of ~30° C though there are many that have narrower or wider ranges than this. These functional ranges are defined by **cardinal temperature points** (Figure 25-1):

- **Minimum growth temperature** – This is the lowest temperature that will allow some measurable amount of microbial growth. Below this temperature key cellular enzymes are inactive so no detectable growth occurs. This doesn't mean the microbes are dead. Many microbes are capable of producing chemicals that function like anti-freeze to keep ice crystals from forming in their cytoplasm. Others can create endospores that have a very low concentration of free water and are therefore impervious to freeze-killing.

- **Optimum growth temperature** – This is the temperature at which optimal growth occurs. Optimal growth doesn't mean that all the enzymes in the cell are working at their maximal rate.

- **Maximum growth temperature** – This is the highest temperature at which the microbial population will continue to grow. Above this temperature, enzymes are permanently denatured at such a rate as to make growth impossible, and the cells die.

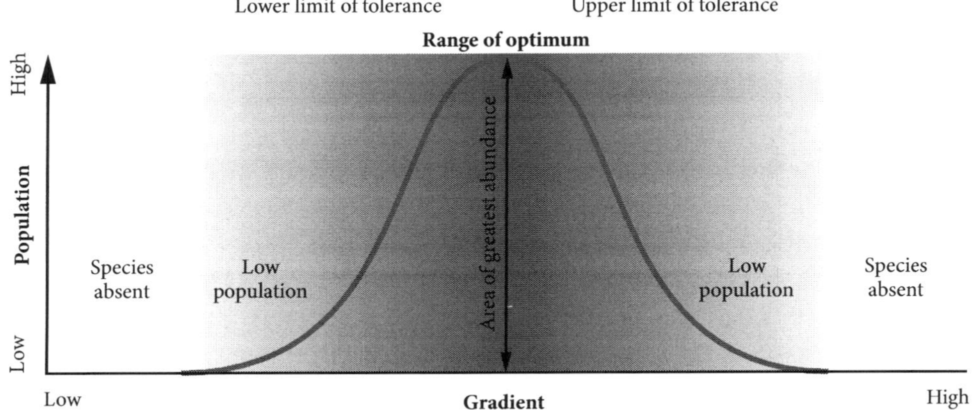

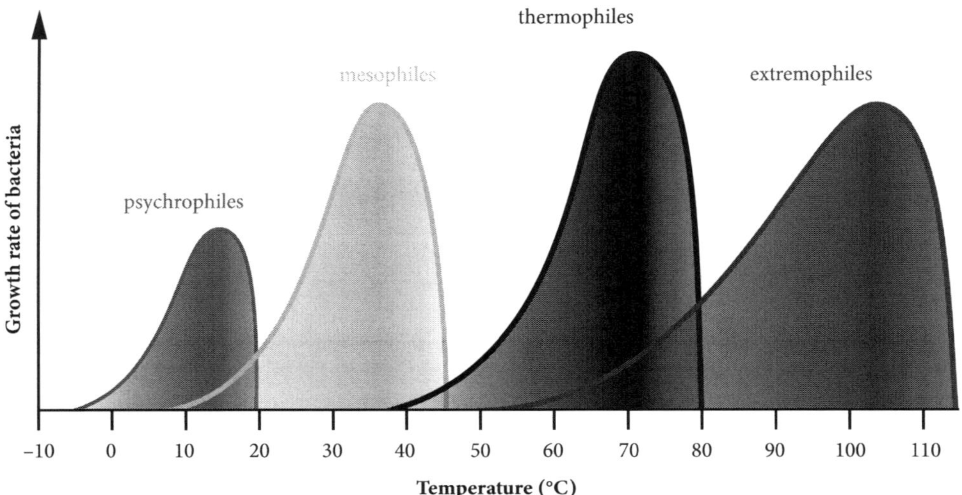

Figure 25-1 | Microbial Growth Cardinal Temperatures

These temperature values can be used to define classes of bacteria. The three basic classifications are as follows:

- **Psychrophiles** – Microbial species that will thrive at ~10° C and generally will grow within a temperature of –5–20° C.

- **Mesophiles** – Microbial species that will grow at human body temperature (37° C) and not above 45° C. Their range is within 20–45° C. They can be split into two subgroups:
  - Those whose optimum lies within 20–30° C, which are usually plant **saprophytes.**
  - Those whose optimum lies within 35–40° C, which are generally organisms that prefer to grow in/on the bodies of **homeothermic** hosts like mammals.

- **Thermophiles** – Microbial species that will grow at 35° C and above. These can be split into two subgroubs:
  - **Facultative thermophiles** – Microbes that will grow at 37° C and with an optimum within 45–60° C.
  - **Obligate thermophiles** – Microbes that will grow only at temperatures in excess of 50° C and whose optimum is above 60° C.

There is a fourth classification category that can be useful: **Extremophiles.** These are microbial species that live and grow on the extreme ends of the environmental conditions found on earth. With regard to temperatures, there are cold extremophiles that will grow at temperatures within the range of –20 to –17° C, and there are hot extremophiles that will grow at temperatures within the range of 100° C to 121° C.

This week, you will be investigating the cardinal temperatures for various bacterial species.

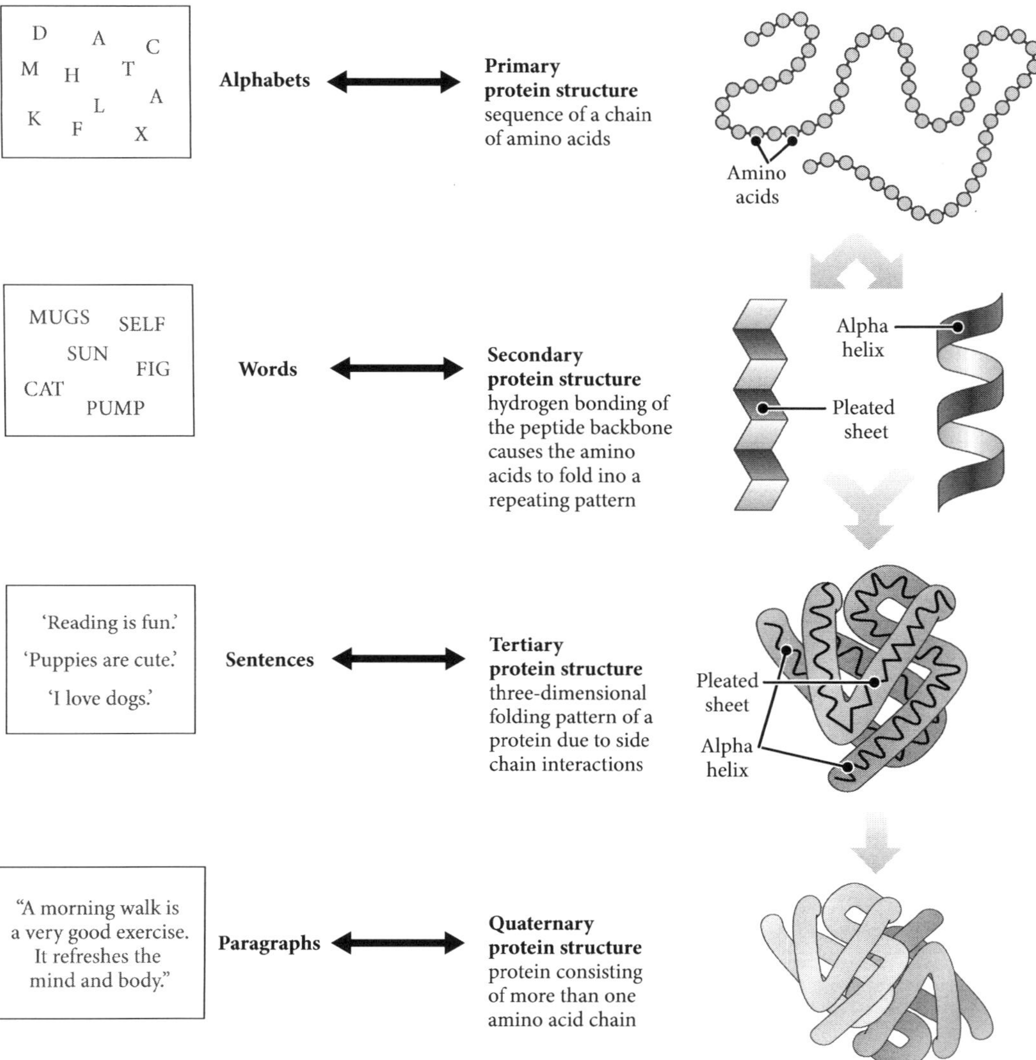

Figure 25-2 | Levels of Enzyme (or General Protein) Structure

Procedure

1 | Obtain cultures of the bacteria and enough tubes of TSB to incubate a sample of each species at the following temperatures: 4° C, 23° C, 37° C, and 60° C.

2 | Mark your tubes with the species and incubation temperature.

3 | Inoculate the TSB with each of the bacterial cultures, being careful to use the same amount of inoculum in each tube.

- Be sure to vortex the stock cultures before you use them. This homogenizes the cell density to the greatest possible degree.

4 | Incubate a sample of each bacterium at each temperature for 24 hrs.

5 | After incubation, vortex each sample and use a pipetter (with tip) to subsample each into a **cuvette.** Fill the cuvette to the fill line to ensure that you have enough subsample to accurately measure it with the **spectrophotometer.**

6 | Be sure that you use a TSB blank to properly calibrate your measurements.

7 | Measure and record the results for each sample in one of the spectrophotometers. Use the data to construct a plot of absorbance or optical density vs. temperature.

Effects of pH on Growth

Temperature is a factor that can affect the growth of microbes, but it isn't the only one. Microbial growth can be affected by pH as well but for different reasons.

pH is the quantitative measure of the acidity or alkalinity of a solution. It is theoretically equivalent to the following:

$$pH = -\log[H^+]$$

Where:

[ ] = concentration

H^+ = hydrogen ion

And:

pH < 7.0 = acidic

pH = 7.0 = neutral

pH > 7.0 = alkaline

Every microorganism has a pH range over which they are able to survive and reproduce, and these typically encompass 2–3 pH units. These organisms also have an optimal pH at which they will grow at their maximal rate. Natural environments are most commonly within the range of pH 3–9, so it shouldn't be surprising that most microorganisms that we have been able to culture have pH optima that lie within this range. As in our investigation of temperature's effects on microbial growth, we can classify microorganisms into three basic categories (Figure 26-1):

- **Acidophiles** – Microorganisms whose pH optimum is below pH 5.5. For these organisms, a high concentration of H^+ is required for the cytoplasmic membrane to remain stable.

- **Neutrophiles** – Microorganisms whose optimum is between pH 5.5–7.9. These are the organisms that could live and grow in/on the human body.

- **Alkalophiles** – Microorganisms whose optimum is pH 8 and up. These organisms are interesting because some of them use Na^+ instead of H^+ to drive materials' transport and motility.

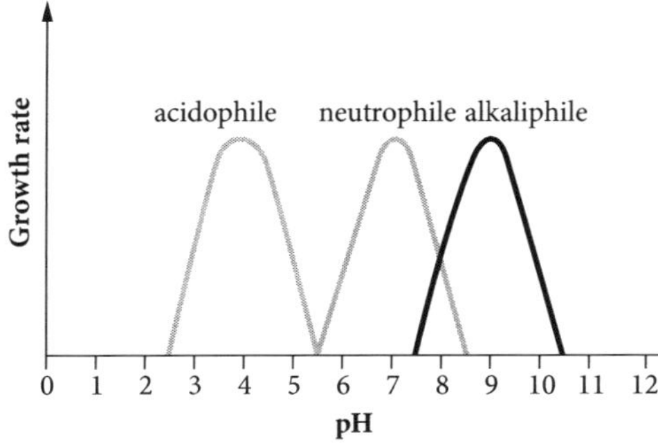

Figure 26-1 | Effects of pH on Growth

In this laboratory exercise, you will be investigating the pH ranges and optima of some bacterial species. You will do this using TSB to which pH **buffers** have been added to raise or lower the pH. Of course, it is worth noting that some of the integral parts of the medium can act as pH-stabilizing buffers such as amino acids, peptones, and proteins. We have added additional chemical buffers to drive the pH of the broth up or down and not merely to make the pH stable.

Procedure

1 | Obtain one tube of each TSB pH for each organism you are investigating. The pHs will be 3.0, 5.0, 7.0, and 9.0.

2 | Inoculate each tube with an equal amount of inoculum.

3 | Incubate the tubes at 37° C for 24 hrs.

4 | After incubation, vortex one of your pH tubes and fill a cuvette with a subsample. Measure the optical density of the subsample using one the spectrophotometers. Now do the same for each of your pH tubes in turn.

Be sure to prepare a blank for each pH broth. Their colors aren't equivalent, and that will yield misleading data if the spec is not properly zeroed.

Laboratory 27
Environmental Osmotic Pressure

A very generalized definition of **osmotic pressure** is "the minimum pressure which needs to be applied to a solution to prevent the inward flow of its pure solvent across a semipermeable membrane" (Voet 2001). To make it particular to our situation, let us change a couple of key words so that we have, "the minimum pressure which needs to be applied to a solution to prevent the inward flow of [water] across [the cytoplasmic] membrane."

In this scheme, **osmosis** is the **net movement** of water molecules across the cytoplasmic membrane from an area of high concentration to low concentration. Concentration of what, again? Water. Water moves from where it is in high concentration to where it is in low concentration. "High" and "low" are relative terms that can only have meaning in relation to something. Higher than what? Lower than what?

One way to think about osmolarity is in terms of tonicity, which is a view from the perspective not of the water (the solvent) but of the solutes (dissolved substances, such as salts). From that perspective, **hypotonic solutions** are those that have a low concentration of solutes and a high concentration of water relative to the solution on the other side of the membrane. **Isotonic solutions** are those that have solute and water concentrations equal to that on the other side. **Hypertonic solutions** have a higher concentration of solutes and lower concentrations of water. So, in all these terms is the implied comparison between the external environment and the internal, cytoplasmic environment.

This **osmotic pressure of the environment** plays an essential role in the survival and reproduction of microorganisms. The cytoplasm of a generalized cell is ~80% water and ~20% dissolved and colloidal substances. For animal cells, which are bounded only by a fragile cellular membrane, a nearly isotonic environment is essential. The range of tolerable osmotic pressures is much wider across the microbial world. In all cases, though, a concentration of solutes and water can be found to induce **plasmolysis** which is the shriveling of cells as water leaves them or **plasmoptysis** which is the swelling of cells as water enters them (Figure 27-1). In animal cells, an acute plasmoptysis results in cell lysis, but in microbes such as bacteria, their cell walls provide a measure of rigidity that can often withstand low osmotic pressures. It is, in fact, common for bacteria to grow best in slightly hypotonic environments in which they maintain a slightly distended, turgid state. It should therefore become clear that various osmotic pressures are not uniformly helpful to all microbes and can therefore be exploited as an antimicrobial agent.

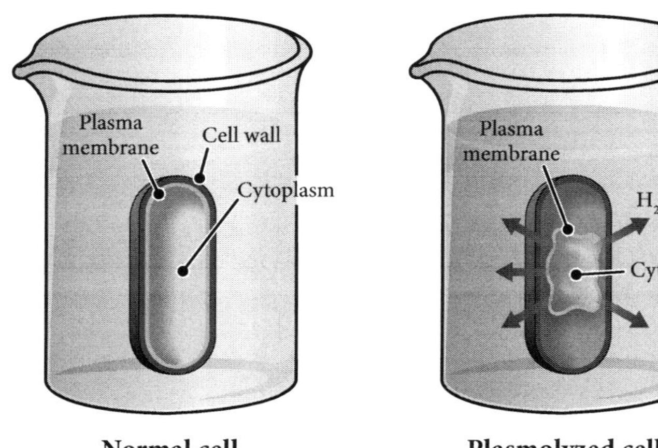

**Normal cell
in isotonic solution**

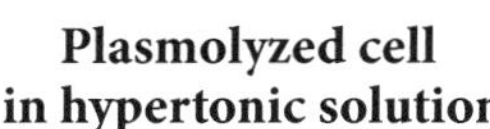

**Plasmolyzed cell
in hypertonic solution**

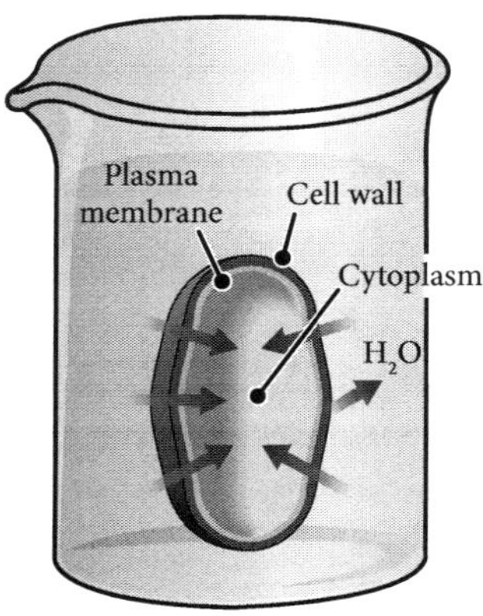

**Plasmoptyzed cell
in hypotonic solution**

Figure 27-1 | The Three Basic Conditions of Osmotic Pressure

Because of the reasons already listed, hypotonic solutions are not very effective as antimicrobial agents. Hypertonic solutions (high osmotic pressure) are fairly effective and are commonly used to inhibit microbial growth. Effective concentrations are not uniform because, as we have seen, microbes exhibit a tremendous diversity. If we use table salt (NaCl) as our model solute, we find that most microbes live and adjust to some range within 0.5% and 3.0% NaCl though there are plenty that can tolerate a bit higher. Concentrations of 10%–15% will completely inhibit the growth of most microbes, and those that are not completely inhibited are deemed **halophiles.**

The purpose of this laboratory exercise is to investigate the effects of different osmotic pressures on the growth of several bacterial species.

Procedure

1 | Obtain cultures of the relevant bacterial species and a TSB tube of each NaCl concentration.

2 | The growth media is a standard TSB to which has been added 0%, 2%, 4%, 6%, and 10% NaCl.

3 | Inoculate a tube of each NaCl concentration with each bacterium.

4 | Incubate at 37° C for 24 hrs.

5 | After incubation, vortex each culture tube and use the contents of each tube to fill a cuvette.

6 | Measure the optical density of each cuvette using the spectrophotometer.

7 | Plot the data for inclusion in the write-up.

References

Voet, Voet, & Pratt (2001). Fundamentals of Biochemistry (Rev. ed.). New York: Wiley. p. 30. ISBN 978-0-471-41759-0.

28

Laboratory 28
Effects of Oxygen Availability

Microorganisms can survive and reproduce over a very wide range of **free/molecular oxygen (O_2)** concentrations. Much like we have discussed when we investigated the effects of temperature and pH, we can use the oxygen requirements microorganisms to classify them in a meaningful way.

A microorganism's oxygen (O_2) requirement is determined by that organism's compliment of bio-oxidative enzymes. This is what gives us a basis for five basic classifications (Figure 28-1):

- **Aerobes** – Require the presence of atmospheric concentrations of oxygen (O_2) for them to grow. They require it because their respiration systems use O_2 as their final e⁻ accepter.

- **Microaerophiles** – Require O_2 but at concentrations lower than those in the typical atmosphere. If they are exposed to environments with atmospheric O_2 concentrations, the excess O_2 blocks the activity of their oxidative enzymes, killing the organism.

- **Facultative** – Also called "facultative anaerobes". Can be thought of as a subset of aerobes that are capable of respiring in O_2-low or O_2-free environments. These organisms will typically use other inorganic ions such as sulfate (SO_4^{2-}) or nitrate (NO_3^-) if there is no O_2 present, or they may use glycolysis + a fermentation pathway.

- **Anaerobes** – Also called "obligate anerobes," require the absence of O_2 because they do not express the enzymes to mitigate the harmful byproducts of aerobic respiration. If O_2 is present, the organism's biochemical machinery will use it as an e⁻ acceptor and thereby create toxic byproducts such as H_2O_2 and O_2^- with which the cells cannot cope. These organisms use other inorganic molecules as final e⁻ acceptors.

- **Aerotolerant** – Here we have microorganisms that do not use O_2 as a final e⁻ acceptor, and they are not poisoned by the presence of O_2. These organisms generally do not do respiration but rather employ only glycolysis + fermentation. They are also generally capable of producing **catalase** and **superoxide dismutase (SOD),** so the toxic compounds that kill anaerobes will not kill them.

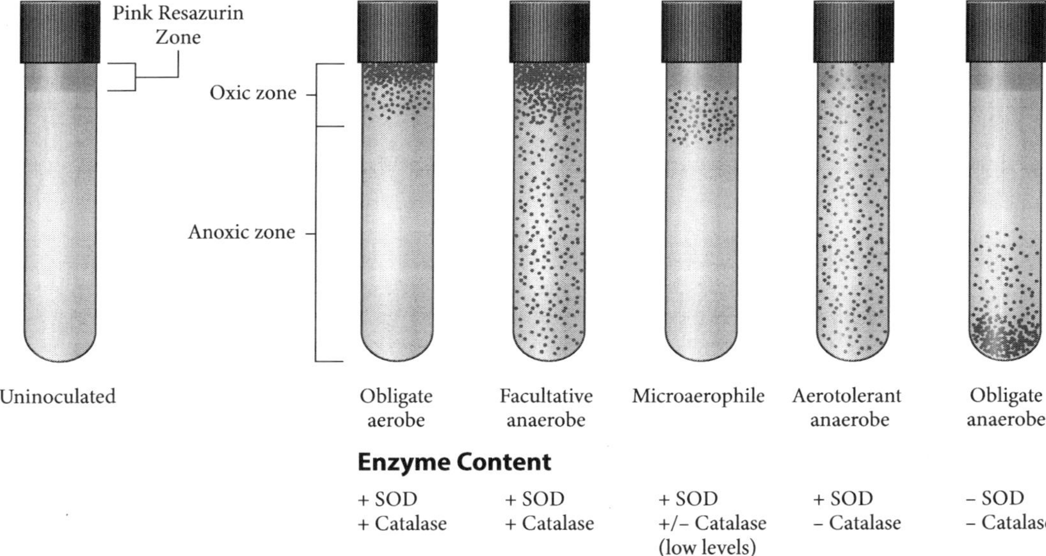

Figure 28-1 | Effects of Oxygen Availability

In this laboratory exercise, you will investigate the oxygen requirements and abilities of a number of microorganisms. To effectively do that, you will need to use a medium that allows the formation of an **oxygen gradient.** The medium that you will use is fluid (semi-solid) thioglycollate (FTG) in which the sodium thioglycollate, thioglycollic acid, and L-cystine reduce the free oxygen in the medium to water. **Autoclaving** drives most of free O_2 out of the medium then the chemical constituents and the semi-solid consistency keep it low, at least for a period of time. The medium also contains **resazurin** which is pink in the presence of O_2 and colorless where O_2 is absent. Finally, the medium is semi-solid, so if you are careful in how you inoculate it, you will be able to determine whether your organism is motile.

Procedure

1 | Obtain a tube of FTG for each organism you are investigating.

2 | Inoculate each tube with a single organism using a standard stab inoculation. That is, don't swirl the inoculation loop or stir the medium. Leave the medium as undisturbed as possible by stabbing directly downward, not touching the bottom of the tube, then remove the loop directly out along the line of inoculation.

3 | Incubate at 37° C for 24 hours.

4 | After incubation, observe the tubes with an eye toward:

    a | Whether there is any pink hue anywhere in the tube.

    b | At what depth(s) did the organisms grow?

    c | Did the organism stay in the inoculation zone, or did any of them begin to migrate outward toward the tube wall?

Bacterial Viruses

Viruses are noncellular infectious agents that replicate only inside of living cells. All viruses are composed, at minimum, of proteins and a single type of nucleic acid. This places them in the chemical category of **nucleoprotein.** They differ from cellular life in notable ways:

1 | They have no measurable metabolism outside of a living cell.

2 | They do not increase in size.

3 | They cannot replicate outside of a living cell whose metabolism they must subvert.

4 | They pass through 0.2 μm filters which are capable of retaining bacteria.

5 | They are almost exclusively **ultramicroscopic,** being viewable only by electron microscopes.

6 | As a completed **virion,** they contain only one type of nucleic acid: DNA or RNA, never both.

Viruses, and indeed several extracellular toxins, were first directly encountered after the advent of filtering technology that was capable of retaining bacteria. Among the first of these devices was the Pasteur-Chamberland porcelain filter that was invented in 1884 (Figures 1 and 2). The first viruses to receive a significant amount of attention were plant pathogens such as tobacco mosaic virus. But even as a large proportion of our understanding of metabolism comes from the study of bacteria, so too does our understanding of viral infection and replication come from the study of **bacteriophages.**

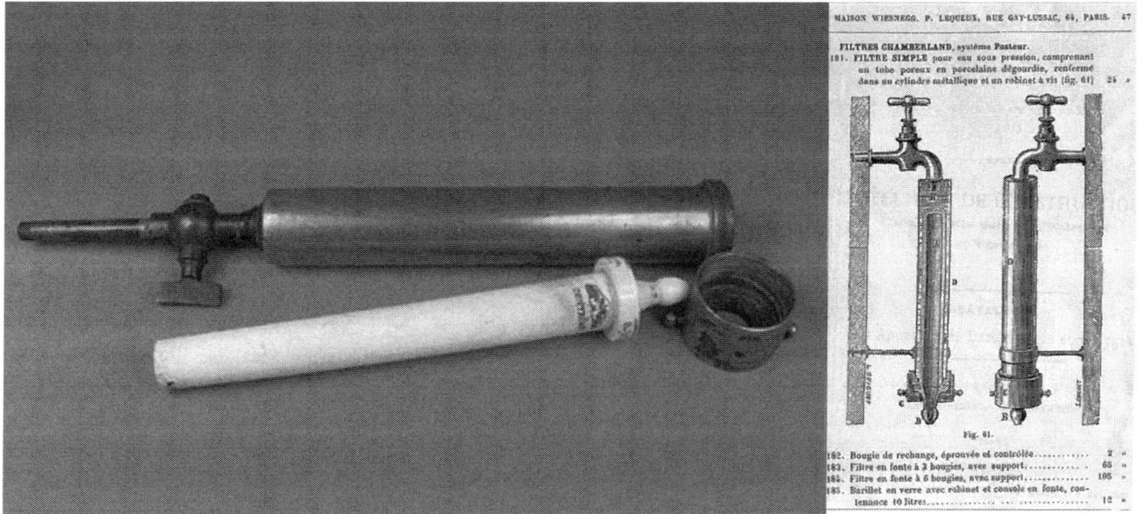

Figure 29-1 | The Pasteur-Chamberland Filter. Composed of a metal body with a porcelain filter insert and designed for use with water pressure.

Left: Courtesy National Museum of American History. **Right:** Courtesy National Library of Medicine.

Figure 29-2 | Another Version of the Pasteur-Chamberland Filter, called the Pasteur Germ Proof Filter. This one works by gravity.

Left: [CC BY 4.0 (https://creativecommons.org/licenses/by/4.0)] https://commons.wikimedia.org/wiki/File:Pasteur-Chamberland-type_water_filter,_London,_England,_1884_Wellcome_L0058326.jpgSee page for author.   Right: [CC0] https://commons.wikimedia.org/wiki/File:Pasteur_Germ_Proof_Filter,_c._1890,_Pasteur-Chamberland_Filter_Co.,_Dayton,_Ohio_-_Museum_of_Science_and_Industry_(Chicago)_-_DSC06633.JPGDaderot

Bacteriophages, generally double-stranded DNA (dsDNA) viruses, were first described in 1915 by F. W. Twort and Félix d'Herelle who came by their discoveries independently. It was d'Herelle who coined them bacteriophages which is Greek for "to eat bacteria" and he was the first person who apply them therapeutically in 1919. In the decades since this initial work, we have determined that bacteriophages (and viruses in general) are variable in size, shape, and complexity. In this laboratory exercise, you will be using the bacteriophage CbK and its host *Caulobacter crescentus* (strain CB15).

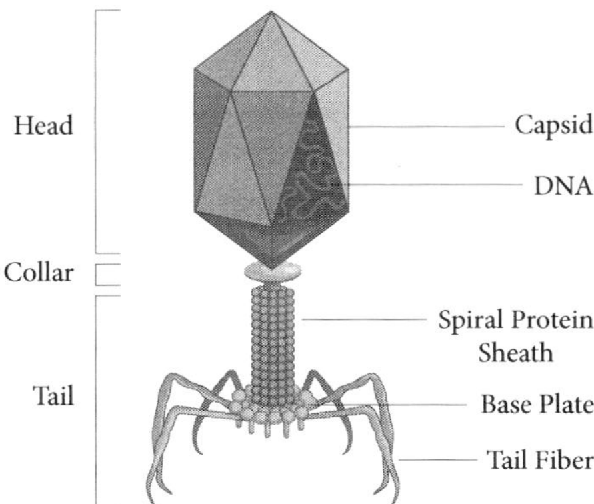

Figure 29-3 | Standard Bacteriophage Anatomy

Phage replication follows a predictable sequence of events:

1 | **Adsorption** – Virus tail fibers reversibly bind to specific carbohydrates exposed on the surface of the cell wall or proteins in the outer membrane.

2 | **Penetration** – Here is where infection occurs. Reversible binding to the exterior is followed by irreversible binding to protein in the cell membrane. The tail fibers contract, the baseplate (which contains lysozyme) changes conformation, and punctures the outer membrane or cell wall. The spiral sheath retracts and the phage's nucleic acid moves through the sheath and into the cell through the pore created by the lysozyme. The empty protein body remains attached to the outside of the cell.

3 | **Replication** – If the phage genome avoids being degraded by enzymes in the host cytoplasm it will begin to subvert the cell's biochemical synthetic machinery and the cell will begin to produce phage components.

4 | **Maturation** – The newly-produced phage components are assembled into complete, or mature virus particles.

5 | **Release** – Lysozyme is produced and breaks down the cell wall, causing the cell to lyse and release the infectious phage particles into the environment.

Phages generally interact with a host cell in one of two ways: the **lysogenic cycle** or the **lytic cycle**. In the lysogenic cycle, a phage successfully infects the bacterium but rather than begin replication, the viral genome is transposed into the bacterial chromosome. The bacterium is then dubbed a **prophage** and phage genome is then passed on to all daughter cells during binary fission. The phage genome will stay in the chromosome until some environmental condition causes it to be transposed back out and will begin the lytic cycle, culminating in lysis of the cell and release of phages. **Temperate viruses**, or **lambda (λ) phages**, are lysogenic and **virulent phages** are lytic.

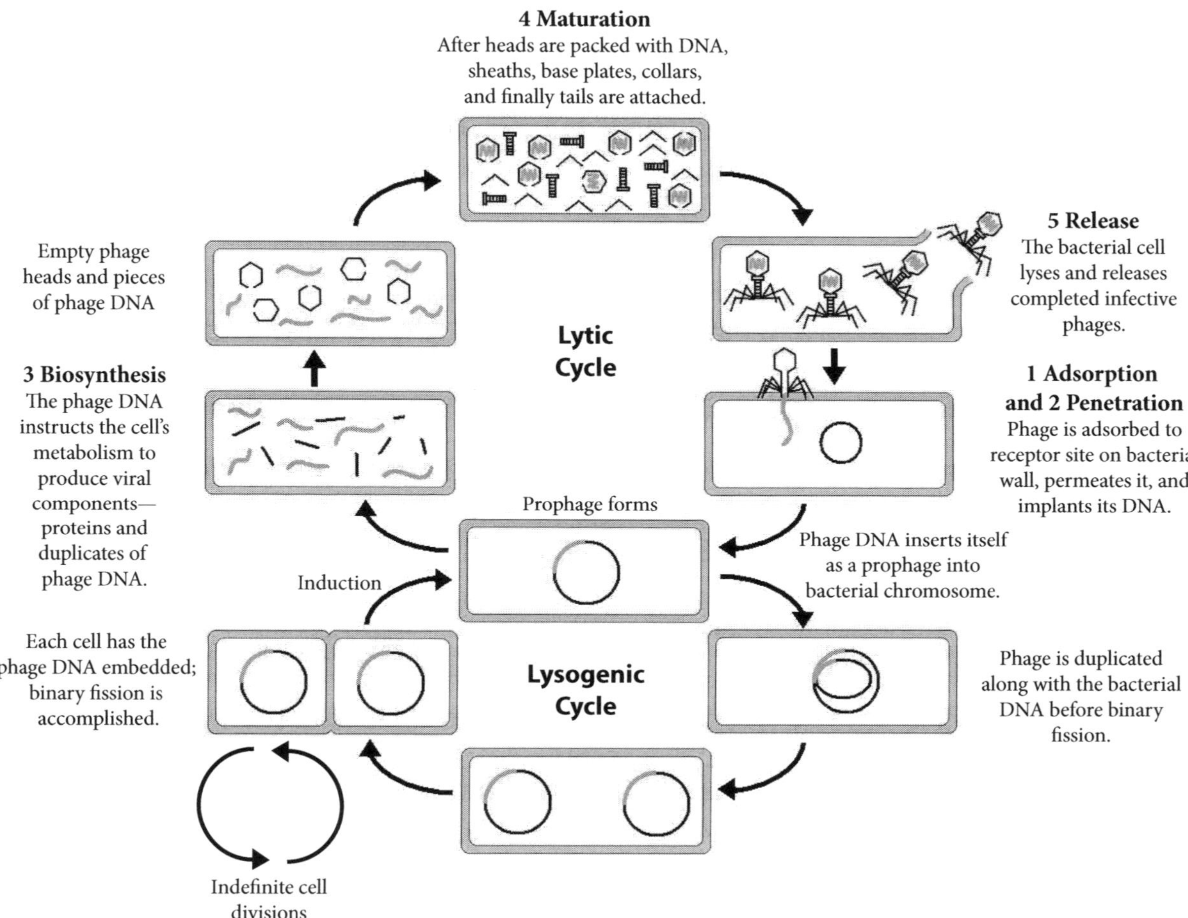

Figure 29-4 | The Lytic (top) and Lysogenic (bottom) Phage Reproductive Cycles

Modified from https://commons.wikimedia.org/wiki/File:Phage2.JPG. Suly12 [CC BY-SA 3.0 (http://creativecommons.org/licenses/by-sa/3.0/)]

While some parallels can be drawn between bacteriophage infection and animal virus infection, they differ in significant ways:

1 | Animal viruses are structurally different than phages. They have no spiral protein sheath, baseplate, or tail fibers, for instance.

2 | Animal virus capsid shapes can be spherical, cuboidal, or helical.

3 | While some animal viruses are dubbed **naked viruses** and are more similar to phages, others are called **enveloped viruses** because they sport a phospholipid membrane outside their capsid. They usually steal the membrane from the cells in which they are replicating and out of which they bud.

4 | Adsorption of animal viruses is accomplished via spikes rather than tail fibers and their receptor sites are on the cytoplasmic membrane rather than a cell wall.

5 | Animal viruses penetrate cells via endocytosis wherein the entire virus enters the cell, capsid and all.

6 | The uncoating of the nucleic acid occurs inside the cell.

7 | The latent period between infection and release of infectious particles is much longer than with phages. Delays of hours or days are common.

In this laboratory exercise you will determine the number of **plaque forming units (PFUs)** in a viral culture. You will create an embedded bacterial lawn into which phages have been introduced. If there are successful infections by lytic phages clear spots will develop in your plates and these are the **plaques**.

Procedure

1 | Obtain a sample of the bacteriophage CbK and note the dilution factor.

2 | Also obtain a culture tube of CB15 and enough tubes of TSB and TSA plates to cover the dilutions.

3 | Serially dilute your stock viral culture to 10^{-9}.

4 | Obtain from the heat block a single tube of molten TSA and *quickly* add two drops of bacterial culture (may use Pasteur pipettes if available) and 0.1 µl of the stock viral culture.

5 | Without delay, recap the tube and briskly spin it between your palms to fully mix the contents.

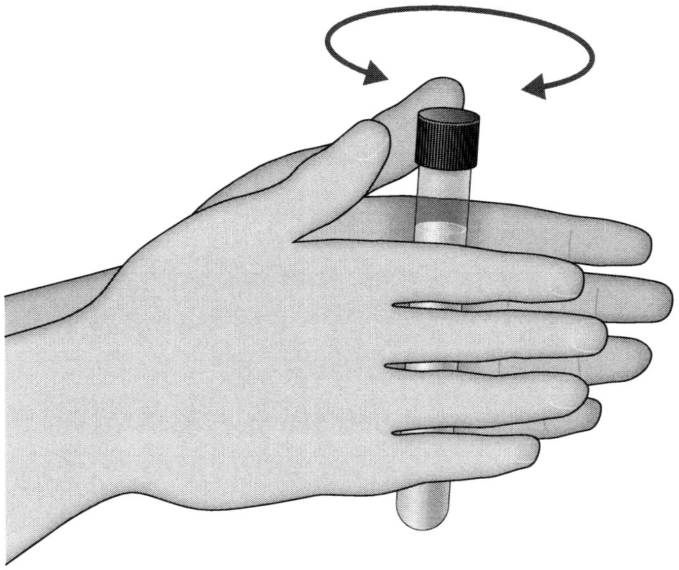

Figure 29-5 | The Proper Way to Mix Bacteria and Bacteriophages in Molten Agar

6 | Immediately pour the molten agar onto the surface of a solidified TSA plate.

7 | Quickly disperse the molten agar evenly over the surface and then STOP MOVING THE PLATE. If you keep moving the plate around you will cause waves to solidify into the agar as it cools. Waves may obscure some of the plaques when you try to count them.

8 | Set the plate aside to fully solidify and repeat these steps with the next dilution factor.

9 | Incubate the plates at 37° C for 24 hours.

10 | After incubation observe the plates and identify one that has 30–300 plaques in it. Count all of the plaques within that plate and convert your count into an estimate of how many PFU were in the original 10^{0} culture.

11 | Plates that have too few or too many plaques should be reported as **TFTC — Too Few To Count** or **TNTC — Too Numerous To Count**.

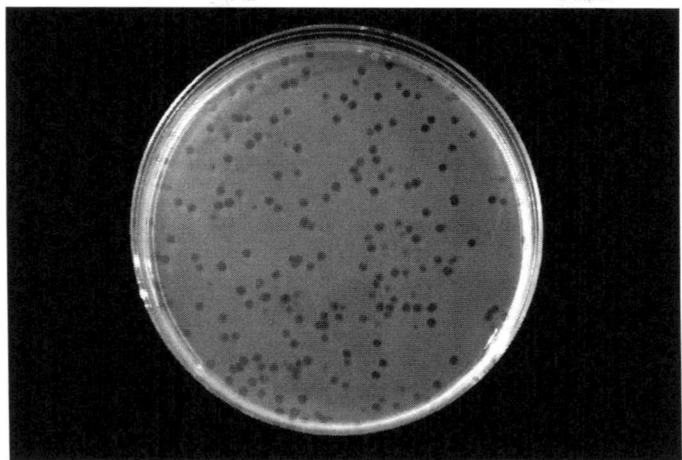

Figure 29-6 | Plaques (the "empty" spots) in a Bacterial Lawn

[CC BY-SA 3.0 (https://creativecommons.org/licenses/by-sa/3.0)] https://commons.wikimedia.org/wiki/File:%CE%A6M12_Plaques_in_Sinorhizobium_meliloti.JPGNinjatacoshell

References

Dupont, Vogensen, Neve, Bresciani, and Josephsen (2004). Identification of the Receptor-Binding Protein in 936-Species Lactococcal Bacteriophages. Applied and Environmental Microbiology, vol. 70 no. 10, pp. 5818-5824.

Rakhuba, Kolomiets, Dey and Novink (2010). Bacteriophage Receptors, Mechanisms of Phage Adsorptionand Penetration into Host Cell. Polish Journal of Microbiology, Vol. 59, No 3, pp. 145-155.

Effects of UV on Cell Viability

Electromagnetic waves (or are they particles?!) exist on a spectrum (Figure 30-1). There are two basic classes of radiation: **ionizing** and **non-ionizing** radiation. Both types consist (on the particle view) of **quanta** with different energies. Ionizing radiation includes things like X-rays and gamma rays which can knock electrons, protons, or neutrons out of molecules of the atoms they pass through. Non-ionizing radiation encompasses everything from ultraviolet light, visible light, infrared light, and up to radio waves. These non-ionizing **wavelengths** may still have certain effects on cells, however. In this laboratory exercise, we are focusing on the **ultraviolet (UV)** section of the spectrum.

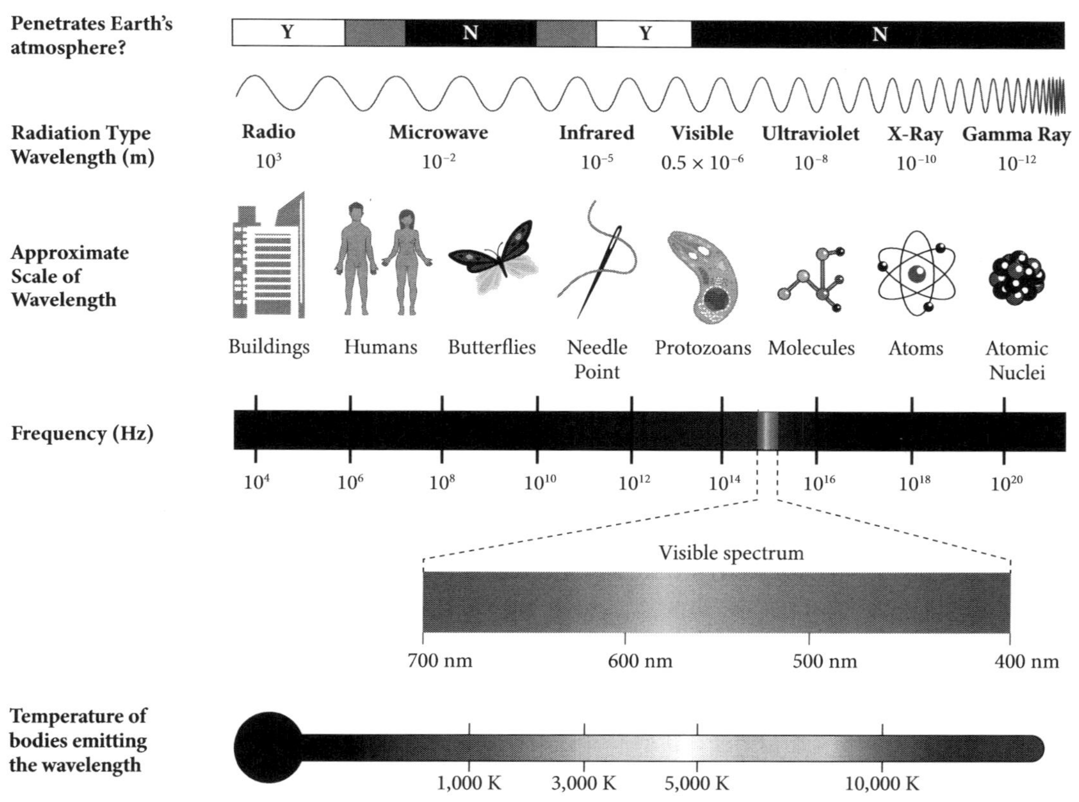

Figure 30-1 | Electromagnetic Spectrum. For a full-color version of this image, see the inside back cover.

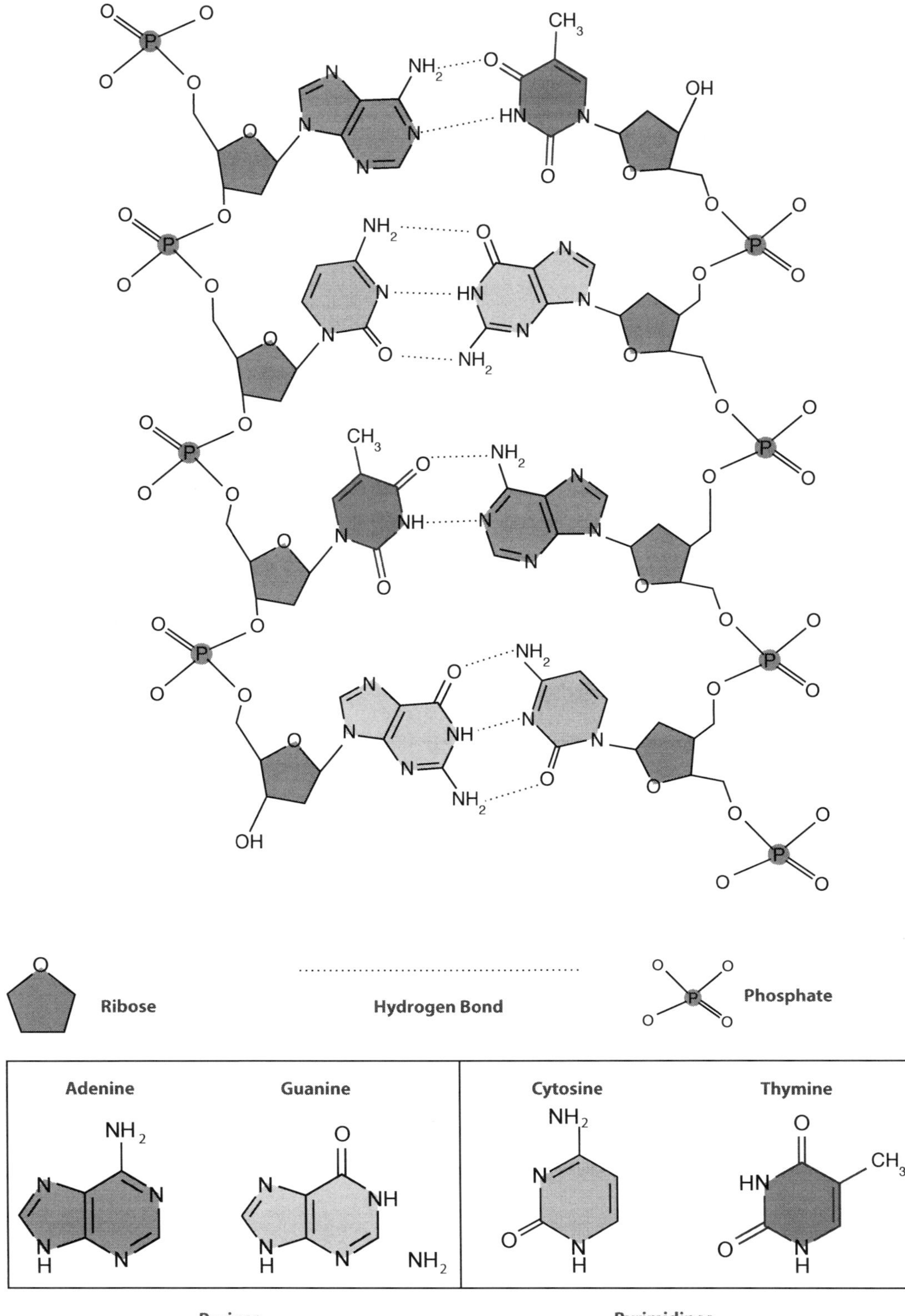

Figure 30-2 | Purines and Pyrimidines

To understand the major impact that UV light may have on living cells, you first must understand the difference between the two basic types of nucleotide: **purines** and **pyrimidines.** Purines include adenine and guanine (A and G), and the pyrimidines include cytosine, thymine, and uracil (C, T, and U). They are classed differently because their nitrogen bases differ (Figure 30-2). Both the purines and the pyrimidines strongly absorb UV light (max for DNA and RNA is ~260 nm). Therefore, the major site of UV cell damage is the pyrimidine nucleic acid bases, due to their chemical structure.

The specific damage done by UV light is in the formation of **pyrimidine dimers** wherein two pyrimidines (C or T) adjacent to one another on the same strand of DNA become covalently bonded to one another. This can impede both the speed of DNA polymerase and its function by greatly increasing the likelihood that it will misread the sequence at the point of the dimer. The greater the density of UV quanta and the longer the exposure duration, the more dimers will be formed.

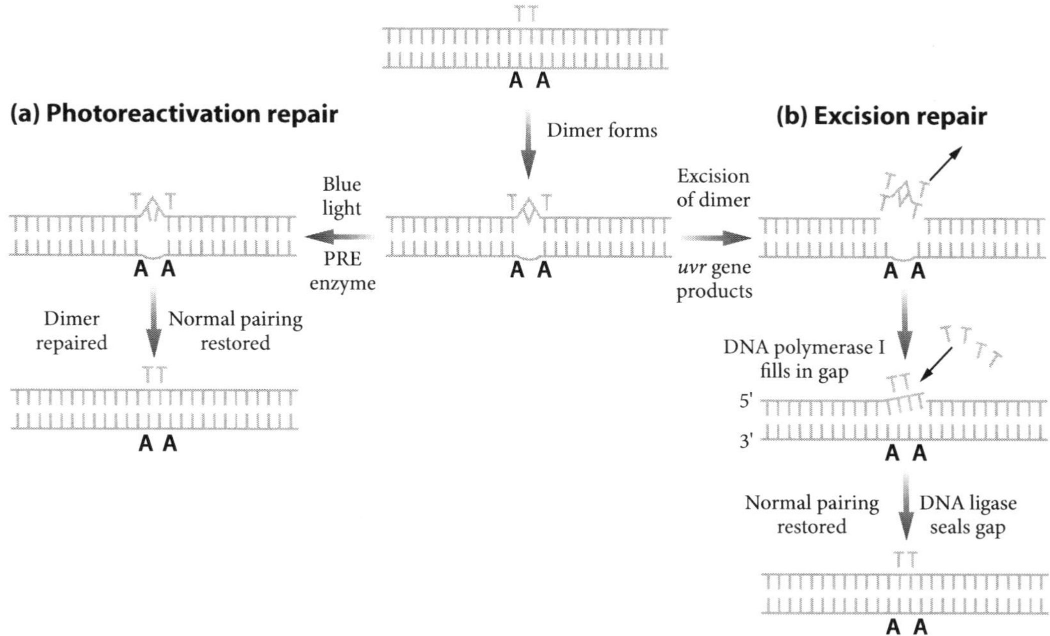

Figure 30-3 | DNA Repair Mechanisms

Microbes have a number of different DNA repair mechanisms that they may use to fix synthesis errors or unlink dimers. Two basic types that can be active when a cell is exposed to UV light are **light repair** and **excision repair** (Figure 30-3). Light repair is activated when a cell has been exposed to light within range of 420–540 nm. This exposure activates a repair enzyme that will split any dimers that were formed. Excision repair is a bit more extreme and does not require light to activate it. Excision repair, as implied in the name, includes detection of a **DNA lesion** (error), **nicking** of the DNA strand containing the error, removal of the offending sequence (or nucleotide), and the writing of a patch using the complementary DNA strand.

In this laboratory exercise, you will be exposing inoculated (but not incubated) plates to UV light for various time intervals. This exposure may or may not impact the viability of the cells on the medium surface, and it is your job to describe what, if any, pattern you see. Finally, it is important to note that UV light has very poor penetrability, so even something as thin as paper or window glass will greatly reduce the amount of UV quanta reaching the cells beneath it. Its use should therefore be limited to the disinfection of **line-of-sight** surfaces and air.

Procedure

1 | Obtain a culture tube of each bacterial species and enough TSA plates to encompass the exposure times and lidded exposure.

2 | Label the plates to indicate that one species will be on one half and the other species will be on the other half.

3 | Inoculate each side of each plate with the given organism using a simple straight-line inoculation.

4 | ***Do not parafilm all the plates.***

5 | Parafilm only the 0 second and lidded exposure plates.

6 | Expose the plates in the UV box, lids off, for 15 seconds, 30 seconds, 1 minute, and 4 minutes. The lidded exposure will have its lid on and should be exposed during the 4-minute interval.

7 | For the health of your eyes and skin, minimize your exposure to the UV light as you are working.

8 | Incubate the plates at 37° C for 24 hrs.

9 | After incubation, observe your plates and note:

   a | Whether the species' responses were the same.

   b | What was the effect of increased exposure duration?

   c | Did the presence of a thin plastic lid have any effect?

   d | Feel free to include a sketch or picture of your results in your write-up.

Evaluating Disinfectants & Antiseptics

The effectiveness of disinfectants and antiseptics does not rely on active metabolism, so they can be used to control viruses, bacteria that produce endospores, regular vegetative bacterial and fungal cells, as well as fungal spores. These compounds are fairly reactive, and they are not selective in their activities so they are limited to use on surfaces (disinfectants), including the outside of organisms (antiseptics). You can find in Table 31-1 a list of some common disinfectants and antiseptics, their modes of action, and usages.

- **Disinfectants** – Are more reactive than antiseptics so their use is generally restricted to non-living surfaces where tissue damage is not a danger.

- **Antiseptics** – Are somewhat less reactive than disinfectants and can be used on the outside of living things. Some antiseptics, like mouthwash, can be used in natural body openings, which is not what many people would think of when they hear the words "external use only".

There are some factors that you need to take into account when you set out to use disinfectants and antiseptics. The manufacturer of these products will often provide some insight into these influential factors the form of usage instructions. In general, you need to consider the following:

- **Concentration** – The concentration of an antimicrobial chemical has a significant impact upon its efficacy. In many cases, more chemical means quicker or more effective lethality, but that is less true for things like isopropyl alcohol which are more effective if they are diluted with a little water. More chemical might also mean tissue or surface damage, so follow the usage instructions and don't be a cowboy (or girl).

- **Duration of exposure** – Chemical reactions are not instantaneous; they actually take some time to proceed. Increasing the duration of the exposure increases the number of chemical reactions that can and do occur.

- **Type of microbe you intend to destroy** – Not every microbe is the same. Trying to disinfect a surface of an enveloped virus like HIV may actually require less work than disinfecting a surface of a spore-forming bacterium or a yeast. Spores (bacterial or fungal), **encapsulated bacteria, acid-fast bacteria,** and naked viruses are more hardy.

- **Environmental conditions**

  - **Temperature** – Temperature controls how fast chemical reactions occur. Raise the temperature and the rate of chemical reactions goes up, lower the temperature and the rate of reaction goes down.

  - **pH** – Some disinfectants are more effective at extreme pHs, and some require an environment fairly close to neutral to prevent the compound from ionizing into a less active form.

  - **Type of material you are treating** – Non-porous surfaces usually require less concentration and less contact time to achieve the same level of **asepsis** as a porous surface.

In this laboratory, you will use a version of the disk diffusion method for evaluating which disinfectants and antiseptics are effective at inhibiting the growth of some bacterial species. Beware during your interpretations that you do not fall into thinking that a larger clear area means more efficacy as it does in the Kirby–Bauer test. We are not controlling for different diffusion rates in this disk diffusion method. Hence, some chemicals may diffuse through the medium more quickly than others which would lead to a larger area of inhibited growth. We are also not adequately testing for the amount of the various compounds contained in each filter disk. What we *can* determine is a bit narrower: We can determine whether any of the compounds are effective. Once we know which are effective, we could do a more targeted test that controls for diffusion rates, such as the use-dilution test.

Procedure

1 | Obtain the bacterial cultures, one TSA plate for each, and a sterile swab for each.

2 | Label each TSA plate with the organism you will inoculate into it.

3 | Make a spread plate with 100 μl of bacterial culture. Be sure to get as even a coverage as you can.

4 | You will then place filter papers that have been wetted with each of various antimicrobial compounds onto the plate.

    a | Label each plate with the compounds you will be testing.

    b | Wet a sterile filter paper using one of the compounds.

    c | Blot the wet filter paper between layers of a paper towel to squeeze out excess liquid. If you leave too much liquid in the paper, a larger amount of compound will diffuse out into the medium and can drown out the results associated with adjacent filter papers.

    d | Using forceps, place a blotted filter paper onto your prepared spread plate in its properly-labelled spot and press firmly down on it with the forceps to ensure it adheres to the medium surface.

    e | Once all the papers are placed, parafilm the plate and leave it face-up for several minutes to ensure the filter papers all finish adhering to the surface. If you fail to wait, your filter papers may fall off the medium surface when they are placed into the incubator in the standard inverted position.

5 | Incubate the plates at 37° C for 24 hrs.

6 | After incubation, observe the plates and note any filter papers with a clear zone around them. These are compounds that were effective and warrant a more targeted test.

Table 31-1 | Common Disinfectants and Antiseptics

Agent	Mode of Action	Use
Phenolic Compounds		
Phenol	Protein denaturation; Precipitation of proteins; Membrane disruption	0.5%–1% solutions antiseptic and mild local anesthetic; 5% solution is disinfectant
Cresols	Similar to phenol	Active ingredient in Lysol; 2%–5% Lysol is disinfectant
Hexachlorophene	Similar to phenol	Topical antiseptic effective against gram+ organisms; Beware of neurotoxic effects
Hexylresorcinol	Similar to phenol	Mild local anaesthetic; Topical treatment of small skin infections, anti-aging creams; Beware of estrogen-like effects
Thymol	Similar to but less effective than phenol	Antifungal; Effective against hookworm; Diluted solutions as mouthwash/gargle

Table 31-1 *continued* | Common Disinfectants and Antiseptics

Agent	Mode of Action	Use
Alcohols		
Alcohols	Lipid solvent; Denaturation/coagulation of proteins; Wetting agent in tinctures; Effect increases with molecular weight	Skin antiseptic; Ethyl 50–70%; Isopropyl 60–70%
Halogens		
Chlorine Compounds	Combination with protein; Forms hypochlorous acid in water; Oxidizer; Allosteric enzyme inhibitor	Sanitizing water; Sanitation of restaurant and dairy utensils; Chloramine 0.1–2% for wound irrigation and dressings
Iodine Compounds	Not fully understood; May precipitate protein; Surface-active agent	Tincture for skin antiseptic; Treatment of goiter; Effective against spores, fungi, and viruses.
Heavy Metals		
Inorganic	Mercuric ion precipitates protein; Allosteric enzyme inhibitor	Beware of irritation and systemic toxicity; Ineffective against spores; Laboratory disinfectant
Organic	Similar to inorganic but more useful	Skin antiseptic; Less toxic, less irritating; Ineffective against spores
Silver nitrate	Precipitation of proteins; Metabolic blocker	Antiseptic of mucous membranes
Surface-Active Agents		
Emulsifiers, soaps, detergents	Lower surface tension of water; Membrane disruption; Alteration of cell permeability; Alter pH of environment	Weak activity against fungi, acid-fast organisms, spores, and viruses
Cationic Agents (quaternary ammonia)	Similar to other surface-active agents; Activity reduced by soaps	Bactericidal, fungicidal; Ineffective against spores and viruses; Asepsis of unbroken skin; Disinfectant for operating-room equipment, dairies, and restaurants
Anionic agents (sodium tetradecyl sulfate)	Similar to other surface-active agents; Activity optimal in acidic conditions	Sclerosing agent in treatment of varicose veins and internal hemorrhoids
Alkylating Agents		
Formaldehyde (gas or liquid)	Adds alkyl groups to nucleic acids and proteins; Crosslinks connective tissues	Active against bacteria, fungi, and viruses; Disinfection of rooms; In alcohol for instrument disinfection; Preserving biological specimens; Deactivation of viruses for vaccine development
Ethylene Oxide	Same as other alkylating agents	Sterilization of materials that would be damaged by heat
B-Propiolactons	Same as other alkylating agents	Sterilization of tissues before grafting; Deactivation of hepatitis viruses; Disinfection of rooms
Crystal Violet	Bonds to nucleic acids and peptidoglycan; Effective against gram+ organisms	Skin antiseptic; Treatment of thrush in infants; Laboratory isolation of gram– bacteria

Laboratory 32
Evaluating Antibiotics

Antibiotics are in a very different class than disinfectants and antiseptics, though they all fall under the broad category of **antimicrobial agents. Antibiotics** are antimicrobial drugs that exhibit **selective toxicity** wherein they are detrimental to a parasite and not to the parasite's host. **Disinfectants** and **antiseptics** are non-selective chemicals that have broad reactivity to a variety of organic compounds.

Antimicrobial compounds work in one or more of these very general ways:

- Interference with cell walls.

- Interference with cell membranes.

- Alteration of the colloidal state of cytoplasm.

- Interference with cellular enzymes.

- Interference with the structure or function of DNA.

- Interference with protein synthesis.

Antibiotics

Antibiotics were originally derived from living microorganisms, mainly **eubacteria, actinomycetes,** and fungi. These organisms possess genes that code for various antimicrobial compounds, and they are capable of expressing those compounds as weapons against other microbes. Many of these microbial warfare compounds exist, but only a small number of them are sufficiently selective in their toxicity so as to allow their use in humans and animals. Unlike disinfectants and antiseptics, antibiotic toxicity is dependent upon there being active cellular metabolism in the targeted organism. A list of common antibiotics, their **modes of action,** and potential side effects is given in Table 32-2.

The **antibiotic resistance** exhibited by some microbes is a topic of great interest and has been since shortly after the widespread use of antibiotic compounds. There are a number of different ways that organisms may exhibit resistance to antibiotics, including the following:

- Efflux pumps that pump out antibiotic compounds before they can harm the cell.

- Expression of enzymes that destroy the antibiotic.

- Alteration of metabolic pathways to exclude susceptible steps.

- Change in the conformation of the antibiotic target, such as an enzyme or ribosome.

Additionally, there are two class of antibiotic-like drugs:

- **Synthetic:** they have no origin in microbes and were constructed wholly in laboratories. These often work as **competitive inhibitors** wherein they slow enzymatic reactions.

- **Semi-synthetic:** the organism that makes the compound must first be fed a synthetic precursor compound that is not found in nature. The microbe modifies the synthetic precursor and produces the final, active product.

In this laboratory, you will use a modified Kirby-Bauer antibiotic sensitivity test to determine how some bacteria respond to six antibiotics. This method controls for the amount of antibiotic (listed on each Sensi-Disc) and for speed of diffusion through the medium. We can therefore use the diameters of any resulting clear zones of inhibited growth to make quantitative determinations about the efficacy of each drug.

Procedure

1 | Obtain the organisms in question and a TSA plate for each.

2 | Prepare a spread plate with 100 µl of bacterial culture. Be sure to get as even a coverage as you can.

3 | Spray a paper towel with a bit of Lysol to make a Lysol wipe.

4 | Also obtain a Sensi-Disc Dispenser and pressing down firmly *once,* dispense the six antibiotic pads. If you hesitate or short-stroke the dispenser, you can cause it to jam.

5 | Wipe the plate-side of the dispenser with your Lysol wipe before passing it to the next person.

6 | Once the Sensi-Discs have been placed on the medium surface, leave the plates face-up for several minutes before parafilming them. If you skip this step, the discs may not fully adhere to the medium surface and fall off when the plates are incubated in the standard inverted position.

7 | Incubate plates at 37° C for 24 hrs.

8 | After incubation, observe your plates, measuring the diameter of any clear zones in mm and compare your data to those in Table 32-3 to determine whether the organism is susceptible, intermediate, or resistant to each antibiotic.

Table 32-1 | Common Antibiotics

Agent	Use	Mode of Action	Side Effects	Spectrum
Ampicillin	UTI, URTI, menigitis, salmonellosis	Penicillin type, wider effectiveness in gram− cells	Rash, nausea, diarrhea, hairy tongue, colitis; Side effects more likely in those with asthma and allergies, especially penicillin allergy.	Broad
Azithromycin	Strep throat, gonorrhea, chlamydia, Legionnaire's disease	Protein synthesis inhibition; binds 50S ribosomal subunit, inhibiting translation	Diarrhea, nausea, abdominal pain, pseudomembranous colitis.	Broad
Chloramphenicol	Eye ointment, plague, cholera, typhoid fever	Inhibits protein synthesis; binds to 50S ribosomal subunit and interrupts protein chain elongation.	Headache, depression, confusion. Aplastic anemia, bone marrow suppression, hypersensitivity reaction.	Broad
Ciprofloxacin	UTI, RTI, typhoid fever, anthrax, chancroid	DNA synthesis inhibition; interferes with DNA gyrase and some topoisomerases.	Nausea, diarrhea, abnormal liver function, vomiting, rash, tendinitis.	Broad, but greater against gram− and less against gram+
Clindamycin	Anaerobic infections, including dental infections	Protein synthesis inhibition; binds 50S ribosomal subunit, inhibiting translation.	Diarrhea, pseudomembranous colitis, nausea, vomiting, abdominal pain/cramps, rash. Metallic taste, vaginal fungal infection.	Aerobic gram+, anaerobic gram−
Erythromycin	Infections of skin and respiratory tract	Not fully understood; binds 50S ribosomal subunit, may prevent the binding of tRNAs.	Diarrhea, pseudomembranous colitis, nausea, vomiting, abdominal pain/cramps, rash. Arrhythmia, anaphalaxis.	Aerobic gram+ and gram−
Gentamicin	Serious infections only	Inhibits protein synthesis; irriversibly binds to 30S ribosomal subunit.	Low blood counts, allergy, neuromuscular anomolies, nerve damage, nephrotoxicity, ototoxicity.	Gram− organisms and Staphylococcus
Methicillin	No longer prescribed	Penicillin type	Formerly used for infections caused by penicillin-G-resistance. No longer manufactured because other, more stable options are available such as cefoxitin.	Gram+ organisms
Penicillin G	Syphilis, strep throat, gas gangrene, leptospirosis and many others	Inhibits bacterial cell wall synthesis by irreversibly binding to DD-tanspeptideases, stopping the final peptidoglycan crosslinking.	Diarrhea, nausea, and hypersensitivity responses including fever, joint pains, urticaria, rashes, angioedema, anaphylaxis	Gram+ plus some gram− outliers in Neisseria and Leptospira
Rifampin	MRSA, Tuberculosis, leprosy, Legionnaire's disease, Lyme disease	Inhibition of RNA synthesis by binding to RNA polymerase and blocking RNA transcript elongation	Hepatotoxicity, breathlessness, flushing, rash, diarrhea, nausea,	Broad
Streptomycin	Plague, tuberculosis, endocarditis	Inhibition of protein synthesis by binding to the 30S ribosomal subunit causing codon misreads	Nephrotoxicity, ototoxicity (specifically to cranial nerve VIII), not to be used during pregnancy but fine while breast feeding.	Broad
Sulfonamides	UTI, bronchitis, bacterial meningitis, ear infections	Slows folate synthesis by competitively inhibiting dihydropteroate synthase	Allergic hypersensitivity	Aerobes
Tetracycline	Acne, cholera, plague, malaria, syphilis	Inhibition of protein synthesis by binding to the 30S ribosomal subunit, blocking peptide elongation	Diarrhea, vomiting, rash, loss of apetite, poor tooth development in those younger than 8 yrs old, nephrotoxicity, increased likelihood of sunburn.	Broad
Vancomycin	Last resort for skin, blood, and endocardial infections	Inhibits bacterial cell wall synthesis by irreversibly binding to un-crosslinked D-alanyl-D-alanine, stopping the final peptidoglycan crosslinking.	Allergic hypersensitivity, nephrotoxicity, ototoxicity, low blood pressure, bone marrow suppression, red man syndrome.	Gram+

Table 32-2 | Results Interpretation Table

Agent	Disk Content	Zone Diameter to the Nearest Whole mm		
		Susceptible	Intermediate	Resistant
Ampicillin				
Gram–	10 µg	≥ 17	14–16	≤ 13
Gram+	10 µg	≥ 17	–	≤ 16
Azithromycin				
Gram–	15 µg	≥ 13	–	≤ 12
Gram+	15 µg	≥ 18	14–17	≤ 13
Carbenicillin				
Chloramphenicol	30 µg	≥ 18	13–17	≤ 12
Ciprofloxacin				
Gram–	5 µg	≥ 31	21–30	≤ 20
Gram+	5 µg	≥ 21	16–20	≤ 15
Clindamycin	2 µg	≥ 21	15–20	≤ 14
Erthyromycin	15 µg	≥ 23	14–22	≤ 13
Gentamicin				
Enterobacteriaceae	10 µg	≥ 15	13–14	≤ 12
Staphylococcus	10 µg	≥ 15	13–14	≤ 12
Methicillin (cefoxitin)				
Enterobacteriaceae	30 µg	≥ 18	15–17	≤ 14
Staphylococcus	30 µg	≥ 25	—	≤ 24
Penicillin G	10 µg	≥ 29	—	≤ 28
Rifampin	5 µg	≥ 20	17–19	≤ 16
Streptomycin	10 µg	≥ 15	12–14	≤ 11
Sulfonamides	250 or 300 µg	≥ 17	13–16	≤ 12
Tetracycline				
Gram–	30 µg	≥ 15	12–14	≤ 11
Gram+	30 µg	≥ 19	15–18	≤ 14
Trimethoprim	5 µg	≥ 16	11–15	≤ 10
Vancomycin	30 µg	≥ 17	15–16	≤ 14

Data Source: Clinical and Laboratory Standards Institute. *Performance Standards Antimicrobial Susceptibility Testing, 28th edition (2018).*

Table 32-3 | Antibiotics Used

Antibiotic	Mass on disk	Label
Streptomycin	10 µg	S 10
Sulfadiazine	0.25 mg	SD .25
Penicillin	10 IU	P 10
Gentamicin	10 µg	GM 10
Vancomycin	30 µg	VA 30
Chloramphenicol	30 µg	C 30

Microbiological Analysis of Food & Water

Humans have been using microbes to preserve and modify foods for as long as humans have been human. Archaeologists have determined that beer was being brewed in ancient China, as many as 5000 years ago (~3000 BC). Microorganisms are a necessary part of many foods today as well. It is yeasts and bacteria that turn grain into beer, milk into yogurt, and smoked ground sausage into pepperoni or salami. Even so, it is equally true that potentially harmful microbes will also grow in/on foods meant for human consumption. A common contaminant in foods that may indicate greater risk are **coliform bacteria,** which are small gram-negative bacilli capable of fermenting lactose. Coliforms are considered an indicator of the possible presence of other dangerous bacteria because they (coliforms) are closely associated with fecal contamination and they are easy to detect. The colons of warm-blooded animals have extremely high concentrations of coliforms (and many other bacteria), so fecal contamination of food or water can be indicated by the presence of coliforms.

Something similar is true with regard to natural waters. There is a wide array of microorganisms that inhabit natural waters and perform essential ecosystem functions. There are **eukaryotic** microbial algae that capture energy from the sun, use it to fix carbon from the atmosphere into complex carbohydrates and proteins, which then serve as food for higher **trophic levels.** There are nitrifying bacteria that convert nitrate (NO_3^-) into nitrite (NO_2^-), ammonia (NH_3), and nitrogen gas (N_2). Some of the same contaminants that you may find in food also act as indicators of fecal contamination in water bodies. When waters have been contaminated by fecal matter, there is a chance that there may be more dangerous species present, such as *Vibrio cholerae,* which is the cause of cholera. Coliforms are an acceptable metric to use in temperate climates where soil temperatures are typically much lower than mammalian body temperatures, but coliform counts will not help you much in warm and wet tropical regions. Here, species like *E. coli* may be found in the soil and water of areas with no provable fecal contamination.

In this laboratory exercise, you will be using one of the most common methods for quantifying viable microorganisms that are associated with food products and natural waters: serial dilution agar plating. You learned this skill in a previous lab exercise, and now you will put your skills to the test using prepared foods purchased from a local deli counter and samples from natural waters in and around Columbia, SC.

Procedure

1 | Obtain six tubes of sterile saline, a food sample, a water sample, six TSA plates, and four each of Eosin Methylene Blue (EMB) and MacConkey (MCA) agar (see page 76 for description).

2 | Label your saline tubes and plates with your name, section number, your food sample, your water sample, and the dilution factors 10^0–10^{-2}.

3 | **Come to an agreement with your laboratory colleagues on the standardized sample size for the various foods.** You may use the available balances to measure out your sample.

4 | Place the analytical fraction of your food sample into the 10^0 saline tube and vortex it to homogenize it as much as possible.

5 | Allow large particles to settle for a few moments and immediately transfer 1 mL of the liquid from the 10^0 tube and place it into the 10^{-1} tube. Vortex the tube 10^{-1} and repeat the procedure again into the 10^{-2} tube.

6 | Once your serial dilutions are done, vortex them one more time to keep any cells suspended in the liquid, then immediately transfer 100 μL into the corresponding TSA and EMB plates and spread them using the glass bar.

7 | Repeat this procedure again for your water sample, using MCA instead of EMB.

8 | Incubate all plates at 37° C for 24 hours.

9 | After incubation, observe the plates looking for visible colonies. If there is a plate that falls between 30–300 colonies, count those colonies and use the count to calculate the number of viable cells and coliforms (if present) in the analytical sample (**number of cells per g or mL**).

10 | Collect results from the other food and water samples for your lab notebook.

1

Assignment 1
Laboratory Notebook

Every experiment that is carried out **<u>must</u>** be typed up in a word document that you will submit to BlackBoard for lab notebook checks. Every experiment will include the following sections: Objective, Procedure, Results, and Interpretations. WE STRONGLY RECOMMEND THAT YOU KEEP A BACKUP COPY OF YOUR DOCUMENT.

The **Objectives** section should be a brief and specific description of the purpose of the experiment and must be written in full sentences.

The **Procedure** section is a description of the experiment as it was carried out in lab and <u>must</u> be written in the past tense. Procedures need to be detailed enough to be able to replicate the experiment. If a procedure is repeated throughout the semester, it has to be written out the first time it is performed, and then you can cite back to your own written procedure. This section should be written in bullet points. The **organism(s) used** in the experiment should immediately follow the procedure. If it is the first time you are working with that particular organism, you must write out the full scientific name (full genus and species). After you have worked with it, subsequent labs can be abbreviated to the genus initial followed by the species. Ex: *Staphylococcus aureus* vs. *S. aureus*

The **Results** section is where you state the observed results of the experiment and it can take the form of a table, a list etc.; whatever form makes it easiest to display them. The results are the <u>visual observations</u> you make about the experiment. You can embed any pictures you want to take as well.

The **Interpretations** section is where you fully explain what your results mean– were your objectives met? What conclusions can you draw based on your results? Discuss whether or not your results were correct and if they were not, explain why. What does this tell you about the organism? How might this result aid it in the environment/clinical setting? In this section the theory behind the experiment is to be used to make logical conclusions. **This section must be written in paragraph form.**

Below is a list of some general comments on typing up your lab notebook.

1 | You will continue to edit the same document throughout the entire semester.

2 | Each new experiment should start on its own new page (meaning insert a page break and label the top of the page with the experiment number and the name of the experiment.

3 | The first page of the document should have your name and lab section and will serve as your table of contents. As you add a new experiment, you will add the experiment name and number to the table of contents along with the page number(s) that correspond to that experiment.

4 | Scientific names are italicized when we type, and underlined when we handwrite. All scientific names in the lab notebook will be italicized.

5 | If you are abbreviating something you must write out the full name or term at least once BEFORE abbreviating, e.g. Tryptic soy agar (TSA).

How Notebooks Will Be Graded

For many of the experiments performed in microbiology, the results are taken the following lab period. This is due to the fact we are working with bacterial cultures that will need time to incubate and grow. This means the first lab's experiments won't have results taken until lab period two. Therefore, the first notebook check upload won't due until the start of the third lab meeting.

There will be 3 notebook check uploads during the semester. Each notebook check will be worth 10 points: 1 point for objectives, procedures, results, and interpretations and 2 points for content/completeness/accuracy.

Write in the due dates for notebook check uploads: _____________, _____________, and _____________.

Example of a Lab Write-Up

Experiment 1. Creating Silence in the Laboratory

OBJECTIVES

The objective of this lab was to determine the longest time period that an entire class of undergraduate students could be silent. This was done to determine whether or not the undergraduate students are able to remain silent whilst their instructor (Teaching Assistant) gives the days' lesson.

PROCEDURE

- Undergraduate students were made to sit in silence and the length of time that the students remained quiet was recorded using a timer.

- This was repeated 3 times throughout the 3-hour class period.

ORGANISMS USED

Homo sapiens (undergraduate students)

RESULTS

The organisms were silent for time periods of 10 minutes, 50 minutes and 120 minutes.

INTERPRETATIONS

With each attempt, the length of time that the organisms were able to sit in silence increased. This shows that it is possible for the organisms to remain silent while their instructor goes through the days' lesson. The 120-minute time period that the organisms remained silent was accomplished because the majority of the organisms fell asleep.

2

Assignment 2
Review of Primary Literature

There are three basic types of scientific literature: **primary, secondary,** and **tertiary.** Primary literature is what you find in scientific periodicals (journals) where the results under discussion are being documented by the individuals who conducted the experiments. Secondary literature is something you may find in textbooks or review articles wherein the authors are experts in the field, but they are writing about work done by others. Tertiary literature is popular-level literature of the sort you may find on a news site, blog, magazine, or some other publication where the writing is being done by someone who has little or no experience with the subject matter and no first-hand knowledge of the experiments that yielded the results they are reporting.

Primary literature is the life blood of scientific and clinical fields. It is in the primary literature that new discoveries are first reported and countervailing hypotheses are tested. The ability to read, understand, and summarize primary literature is therefore a key skill for students planning to enter a clinical field.

Assignment

This assignment will be due on ____.

You will read and summarize one of the articles on Blackboard under course documents. Your summary will be in Times New Roman, 12-pt font, single-spaced and will be 1–1.5 pages in length. Normal 1" margins. *No* huge "chunks" of paragraphs with direct quotes from the article. This is a summary; every word/sentence you use must be your own.

At least three full paragraphs for the summary (think: introduction of the topic, results, and a discussion paragraph). You may go into some procedural details if you find it necessary. The last paragraph of your summary will be your opinion on the article. Answer the following questions to help you:

- Did you like/not like this article? *Why?* (The why is the most important part; give specific details on why you did or didn't like this article.)

- What could the authors have improved on/made clearer?

- In your opinion, *why* is this research important/unimportant?

- How could this relate to the field/career you plan to pursue?

This will be uploaded to Safe Assign on blackboard, as well as turned in during lab time as a hard copy.

Your summary must include at least one in-text article citation (from the one you are reviewing!) with an accompanying full citation after the end of the summary portion. These in-text and full citations should comport with The Council of Science Editors (CSE) format.

General Outline of the CSE Format Used in This Lab

Author (Publication Year). Article Title. Journal Title; Volume(Issue): Page Numbers.

Koronakis, V., Cross, M., Senior, B., Koronakis, E., and Hughes, C. (1987). The secreted hemolysins of *Proteus mirabilis, Proteus vulgaris,* and *Morganella morganii* are genetically related to each other and to the alpha-hemolysin of *Escherichia coli.* Journal of Bacteriology; April, 169(4): 1509–1515.

Life Pro Tips

1 | Whenever you come across a term you aren't familiar with, plug it into *Dictionary.com* or just Google it.

2 | Read the primary article several times to increase your understanding of the information presented.

3 | Form an outline of the article before writing a rough draft to collect your thoughts and to form a firm understanding of what you are writing.

4 | Revise your rough draft and reread your summary in its entirety aloud to account for awkward sentence structures, typos, and misinformation.

5 | Make sure your paper was properly submitted to Safe Assign in Blackboard. If you are unsure, email the TA before the assignment deadline.

3

Assignment 4
24 Hour Return Instructions

1 | **Get a wood block and pack all your tubes into it. You have two VP tests. This test takes the longest. Start both first.**

2 | **EnteroPluri**

    a | Ask your TA to open up compartments 4 (H_2S/Ind) and 9 (VP).

    b | To compartment 4 (H_2S/Ind), make sure you check for the result for H_2S first. It will appear black around the inoculation needle in the tube. Then add 2–3 drops of Kovac's reagent (available on TA desk) and observe the **reagent layer** (not the agar) for a color change to red for a positive test. This will happen within a few seconds.

    c | To compartment 9 (VP) add 2–3 drops of **Barritt's A Reagent** followed by 2–3 drops of **Barritt's B Reagent.** Wait 15–20 minutes and check for a deep red color to develop at the top of the medium.

        ■ **CAREFUL WITH THE ENTEROTUBE AFTER YOU ADD REAGENTS!***

        ■ If tilted, the reagents will spill into other compartments, changing the appearance of the results of other test chambers.

    d | Using your the instructions in Laboratory 17 check all 12 compartments for color changes. Comparison sheets are available.

    e | Fill out the results sheet to determine an ID number. Bring results sheet to your TA and they will give you the EnteroPluri Test ID booklet used to determine the organism identity associated with your ID number.

    f | Check the results of all your other tests while you're waiting for your VP tests!

3 | **IMViC VP TEST:** Obtain an empty, sterile test tube and pour roughly half of the contents of your MR-VP into it. Label one tube MR, the other VP

    ■ Add 5 drops of each Barritt's A and B Reagent, mixing between additions, to one of the tubes and label that tube "VP".

    ■ Check the VP tube when you check your Enterotobe VP test (needs 15–20 minutes to react).

4 | **IMViC MR Test:** Add 3–5 drops of **Methyl Red Reagent** to the MR tube and record the color.

5 | **IMViC Citrate utilization: part of your Enterotube**

6 | **IMViC Indole: part of your Enterotube**

7 | **Carbohydrate fermentation**

    ■ Check the glucose and sucrose broth tubes for color change from red to yellow (indicative of acid production) and possible gas by-product in the Durham tube (liquid medium will be displaced, bubble will appear at the top of the tube).

8 | **Nitrate reduction**

- To the trypticase nitrate broth culture add 5 drops of **Nitrate A** followed by 5 drops of **Nitrate B.**

- Look for a red color change.

- If there is no color change, add a small amount of zinc powder and observe.

9 | **O_2 Tolerance Tube:** Observe the growth in the tube. Is it growing just in the oxic zone? Just in the anoxic? Equally throughout? Better in the oxic but still growing some in the anoxic? Based on the growth, classify its O_2 tolerance using the growth categories in Laboratory 28.

10 | **Has it been 20 minutes?** Check your VP tests and record the results.

11 | Remove tube labels and rack them into the wire baskets, EnteroPluri tubes go into solid biohazard waste.

12 | Wash your hands before you leave.

4

Assignment 3
Lab Report on Unknowns

This Is Due

The lab report will contain an Introduction, Materials and Methods, Results, Discussion, and Reference section.

When writing the report the voice should be consistent (*i.e.* 'a catalase test was performed on the unknown bacterium'). When describing the procedures the past tense should be used. The report and theory however, is to be written in the present tense (*i.e.* 'The purpose of this report is' or 'the catalase test distinguishes between organisms').

A good lab report does more than just present a laundry list of data; it demonstrates the writer's comprehension of the concepts behind the data and makes a case for why the data support your identification.

Lab reports must be uploaded into SafeAssign on Blackboard by the beginning of scheduled lab time on April 10th. Failure to turn in a hard copy of your lab report/upload your report into SafeAssign will result in *NO CREDIT* awarded for the assignment. Reports containing plagiarized sections will result in a *failing grade of ZERO (0)* for the assignment and a referral to the Office of Academic Integrity.

If you are a repeating student, your unknown report must include significant revisions from the previous version to acquire a SafeAssign assessment that does not indicate plagiarism. Consult with your TA if you have questions about this.

Formatting

- Do not include a cover page
- Name and section number should be at the top of the page (no additional header) and then a title of the report
- Lab reports will be typed in size 12 pt Times New Roman font
- Regular 1" margins should be used.
- Materials and methods section **must be single spaced.** Everything else can be up to 1.5 spacing, no double spacing.
- Must print front and back
- Staple Rubric to the front or back of your hard copy

Introduction

- States the overall big picture objective of the experiment and gives a background to the experiment. Why is it important to identify unknown bacteria? How does this relate to real world application?
- Introduce your unknown bacterium. (Basic facts, secondary sources) Note: Wikipedia will not be accepted as a source; for secondary sources try .gov websites like the CDC of FDA website.
    - Who cares about the bacterium you are talking about?
    - What is interesting about your unknown?

- **You will perform a primary literature search on your Unknown Organism.** This means reading primary literature (i.e. peer reviewed research journal articles) and report on your organism. This should be more than just one paragraph or one citation. You actually need to do some research! What research is being done on this organism? What role does it play in a clinical/environmental setting, etc.? **For your primary literature article, follow closely to how you did your review of the article for the literature assignment. You CANNOT use one of the articles from the previous literature assignment. If you use one of those, it will result in ZERO points for that portion.**

- **You also need to review a primary literature article that's about identifying unknown organisms.** It can be a clinical case study that involved making an unknown bacterial identification, or a research article about the ID'ing of an unknown organism, but it has to be a primary article. It does not have to be on your unknown bacterium specifically! If you are unsure if it's primary, ask! Primary literature articles that discuss a protocol or techniques used to ID microbes does NOT count!

- **REVIEW ARTICLES ARE NOT CONSIDERED PRIMARY LITERATURE RESOURCES. Hint: If it says, "review article" or "minireview", it is a review article.**

Methods and Materials

- This section should be a description of all the tests performed on the unknown bacterium. Each test should have its own heading followed by a brief description (a few sentences) of the test procedure and **objective. (Make sure to include the reagents that are used in that particular experiment).** If abbreviations are used the full description must be given before the abbreviation. Make sure you cover the objective of each experiment clearly. I do not need an exact step-by-step procedure of what you did for each experiment. I need what the test is looking for and the **overall objective.** Include what a positive and negative result looks like, and each result means.

e.g. ***Carbohydrate fermentation***

The ability of the unknown bacterium to ferment carbohydrates to an acidic end was tested by inoculating it in sucrose and glucose broths containing inverted Durham tubes. The inoculated media were incubated at 37°C for 24 hrs. A positive result will show a color change in the medium from red to yellow if the organism can ferment the carb to an acidic end. The Durham tube will trap any gas that is produced during fermentation. A negative result will produce no color change to the medium, meaning it will stay red.

Results

This section must include a table (or several tables) with **ALL** the test results (i.e. observations, color changes, etc.) for the unknown bacterium with **basic interpretations.**

Example of a results table:

Table 4-1 | Results of Tests Carried out on the Unknown Bacterium

Test	Results and Interpretation
Appearance on Solid Media	Small, yellow, circular colonies with entire margin and raised elevation
Gram Stain	Pink rod-shaped cells in chains; Gram negative streptobacilli
Methyl Red Test	Red color change. Positive for glucose fermentation to an acidic end

Taxonomy

- You need to identify the unknown bacterium down to species. Give its most current taxonomic hierarchy/lineage—a good source is the NCBI Taxonomy browser.

  http://www.ncbi.nlm.nih.gov/taxonomy

  e.g., Escherichia coli

  Superkingdom: *Bacteria*

  Phylum: *Proteobacteria*

  Class: *Gammaproteobacteria*

  Order: *Enterobacteriales*

  Family: *Enterobacteriaceae*

  Genus: *Escherichia*

Discussion

- In this section you need to analyze and interpret the results as well as discuss how you arrived at the concluded identification. Ask yourself the following questions:

  - What do the results clearly indicate? How did the test results and the physiological and biochemical characteristics of the unknown lead to the concluded identification?

  - What characteristics of the unknown bacterium distinguish it from all other organisms other than its identified species?

  - Was there an experiment(s) that was particularly useful for identifying your unknown?

- The table on pg. (insert manual page #) of the lab manual may be mentioned in the paper but needs to be checked and confirmed with other credible sources. Stating that you were able to determine the identity of the unknown based on an elimination process using the chart on page (insert manual page #) is not a suitable explanation all by itself.

- **Were there any conflicting results?** Why or why did not this happen? (the answer is not always human error!) Analyze any experimental error; contradictory results must be explained.

- **Important: The discussion is NOT just a regurgitation of your results table! The discussion is where you explain how you used your results to get to your unknown ID.**

- Your last paragraph of your discussion should be a concluding paragraph that sums up the project: re-state the major purposes, important organism details, and the identity of the unknown bacterium.

  - What did you learn from this?

  - How can what you learned be applied to a clinical setting or what you want to do professionally after undergrad, etc.?

Works Cited

List *any & all* sources used, *i.e.* lab manual, websites, journal articles, and books. Scientific format must be used and in text citations are expected with the use of external sources. **CSE format** only.

General Format Guide and Examples

Author(s). Publication Year. Article Title. Journal Title; Volume(Issue): Page numbers.

Sugiyama, T., Iida, T., Izutsu, K., Park, K.S., Honda, T. 2008. Precise region and the character of the pathogenicity island in clinical *Vibrio parahaemolyticus* strains. Journal of Bacteriology; 190(5): 1835-1837.

Hagenbuch, I. (2019). Microbiology Lab, BIOL 250L. Van-Griner, p. 157.

Deacon, J. (2010). The Microbial World: *Proteus vulgaris* and clinical diagnostics. Institute of Cell and Molecular Biology, The University of Edinburg; http://archive.bio.ed.ac.uk/jdeacon/microbes/proteus.htm (Accessed March 25, 2019).

In-Text Citations:

1 author: (Gartmon, 2018)

2 authors: (Gartmon and Lovell, 2018)

3+ authors: (Gartmon *et al.,* 2018) (note the period after "*al*")

If using a .gov or website without authors, use the company name and the year the webpage was last edited

Examples: (CDC, 2018), (Microbe Wiki, 2014), (FDA, 2011)

Grading Rubric for Lab Report on Unknown

Name: ____

Criteria	Points Possible	Points Earned
Introduction ▪Content knowledge is accurate and relevant; demonstrates a clear understanding of the big picture ▪**Introduce objective of project (3 pts)** ▪**Introduce basic facts about unknown (5 pts)** ▪**Primary Lit review on unknown (5 pts)** ▪**Case Study ID'ing Unknown bacteria (5 pts)**	18	
Materials and Methods ▪Summaries are clear, concise description of experiments ▪Objectives, reagents, +/– results (meaning what they look like) and +/– interpretations (what the results mean)	15	
Results Table ▪Data presented in logical format and labeled correctly. Includes both the experiment result and brief interpretation	10	
Taxonomy ▪Unknown organism ID is correct/taxonomic hierarchy is correct	5	
Discussion ▪Clear logical explanations of how results led to identification, using other sources to confirm results (7.5 pts total, 2.5 pts comes from outside sources, 5 pts logical explanation) ▪Explain clearly conflicting results/why you got what you got (not just stating human error. **Critical thinking.** (5 pts) ▪Concluding paragraph (2.5 pts)	15	
References ▪Bibliography done in scientific format; citations included in text	5	
Writing quality ▪Clear, concise, direct ▪Grammar, spelling, formatting	2	
Total Points	70	